I, Mathematician

© 2015 by
the Mathematical Association of America, Inc.

Library of Congress Catalog Card Number 2015932464

Print edition ISBN: 978-0-88385-585-0

Electronic edition ISBN: 978-1-61444-521-0

Printed in the United States of America

Current Printing (last digit):
10 9 8 7 6 5 4 3 2 1

I, Mathematician

Edited by

Peter Casazza
University of Missouri-Columbia

Steven G. Krantz
Washington University

Randi D. Ruden
University City, Missouri

Published and Distributed by
The Mathematical Association of America

Council on Publications and Communications
Jennifer J. Quinn, *Chair*

Committee on Books
Fernando Gouvêa, *Chair*

Spectrum Editorial Board
Gerald L. Alexanderson, *Co-Editor*
James J. Tattersall, *Co-Editor*

Virginia M. Buchanan   Thomas L. Drucker
Richard K. Guy   Dominic Klyve
Shawnee L. McMurran   Daniel E. Otero
Jean J. Pedersen   Kenneth A. Ross
Marvin Schaefer   Franklin F. Sheehan
Amy E. Shell-Gellasch

SPECTRUM SERIES

The Spectrum Series of the Mathematical Association of America was so named to reflect its purpose: to publish a broad range of books including biographies, accessible expositions of old or new mathematical ideas, reprints and revisions of excellent out-of-print books, popular works, and other monographs of high interest that will appeal to a broad range of readers, including students and teachers of mathematics, mathematical amateurs, and researchers.

777 Mathematical Conversation Starters, by John de Pillis

99 Points of Intersection: Examples—Pictures—Proofs, by Hans Walser. Translated from the original German by Peter Hilton and Jean Pedersen

Aha Gotcha and Aha Insight, by Martin Gardner

All the Math That's Fit to Print, by Keith Devlin

Beautiful Mathematics, by Martin Erickson

Calculus and Its Origins, by David Perkins

Calculus Gems: Brief Lives and Memorable Mathematics, by George F. Simmons

Carl Friedrich Gauss: Titan of Science, by G. Waldo Dunnington, with additional material by Jeremy Gray and Fritz-Egbert Dohse

The Changing Space of Geometry, edited by Chris Pritchard

Circles: A Mathematical View, by Dan Pedoe

Complex Numbers and Geometry, by Liang-shin Hahn

Cryptology, by Albrecht Beutelspacher

The Early Mathematics of Leonhard Euler, by C. Edward Sandifer

The Edge of the Universe: Celebrating 10 Years of Math Horizons, edited by Deanna Haunsperger and Stephen Kennedy

Euler and Modern Science, edited by N. N. Bogolyubov, G. K. Mikhailov, and A. P. Yushkevich. Translated from Russian by Robert Burns.

Euler at 300: An Appreciation, edited by Robert E. Bradley, Lawrence A. D'Antonio, and C. Edward Sandifer

Expeditions in Mathematics, edited by Tatiana Shubin, David F. Hayes, and Gerald L. Alexanderson

Five Hundred Mathematical Challenges, by Edward J. Barbeau, Murray S. Klamkin, and William O. J. Moser

The Genius of Euler: Reflections on his Life and Work, edited by William Dunham

The Golden Section, by Hans Walser. Translated from the original German by Peter Hilton, with the assistance of Jean Pedersen.

The Harmony of the World: 75 Years of Mathematics Magazine, edited by Gerald L. Alexanderson with the assistance of Peter Ross

A Historian Looks Back: The Calculus as Algebra and Selected Writings, by Judith Grabiner

History of Mathematics: Highways and Byways, by Amy Dahan-Dalmédico and Jeanne Peiffer, translated by Sanford Segal

How Euler Did Even More, by C. Edward Sandifer

How Euler Did It, by C. Edward Sandifer

Illustrated Special Relativity Through Its Paradoxes: A Fusion of Linear Algebra, Graphics, and Reality, by John de Pillis and José Wudka

I, Mathematician, edited by Peter Casazza, Steven G. Krantz, and Randi D. Ruden

In the Dark on the Sunny Side: A Memoir of an Out-of-Sight Mathematician, by Larry Baggett

Is Mathematics Inevitable? A Miscellany, edited by Underwood Dudley

I Want to Be a Mathematician, by Paul R. Halmos

Journey into Geometries, by Marta Sved

JULIA: a life in mathematics, by Constance Reid

The Lighter Side of Mathematics: Proceedings of the Eugène Strens Memorial Conference on Recreational Mathematics & Its History, edited by Richard K. Guy and Robert E. Woodrow

Lure of the Integers, by Joe Roberts

Magic Numbers of the Professor, by Owen O'Shea and Underwood Dudley

Magic Tricks, Card Shuffling, and Dynamic Computer Memories: The Mathematics of the Perfect Shuffle, by S. Brent Morris

Martin Gardner's Mathematical Games: The entire collection of his Scientific American columns

The Math Chat Book, by Frank Morgan

Mathematical Adventures for Students and Amateurs, edited by David Hayes and Tatiana Shubin. With the assistance of Gerald L. Alexanderson and Peter Ross

Mathematical Apocrypha, by Steven G. Krantz

Mathematical Apocrypha Redux, by Steven G. Krantz

Mathematical Carnival, by Martin Gardner

Mathematical Circles Vol I: In Mathematical Circles Quadrants I, II, III, IV, by Howard W. Eves

Mathematical Circles Vol II: Mathematical Circles Revisited and Mathematical Circles Squared, by Howard W. Eves

Mathematical Circles Vol III: Mathematical Circles Adieu and Return to Mathematical Circles, by Howard W. Eves

Mathematical Circus, by Martin Gardner

Mathematical Cranks, by Underwood Dudley

Mathematical Evolutions, edited by Abe Shenitzer and John Stillwell

Mathematical Fallacies, Flaws, and Flimflam, by Edward J. Barbeau

Mathematical Magic Show, by Martin Gardner

Mathematical Reminiscences, by Howard Eves

Mathematical Treks: From Surreal Numbers to Magic Circles, by Ivars Peterson

A Mathematician Comes of Age, by Steven G. Krantz

Mathematics: Queen and Servant of Science, by E.T. Bell

Mathematics in Historical Context, by Jeff Suzuki

Memorabilia Mathematica, by Robert Edouard Moritz

More Fallacies, Flaws, and Flimflam, Edward J. Barbeau

Musings of the Masters: An Anthology of Mathematical Reflections, edited by Raymond G. Ayoub

New Mathematical Diversions, by Martin Gardner

Non-Euclidean Geometry, by H. S. M. Coxeter

Numerical Methods That Work, by Forman Acton

Numerology or What Pythagoras Wrought, by Underwood Dudley

Out of the Mouths of Mathematicians, by Rosemary Schmalz

Penrose Tiles to Trapdoor Ciphers ... and the Return of Dr. Matrix, by Martin Gardner

Polyominoes, by George Martin

Power Play, by Edward J. Barbeau

Proof and Other Dilemmas: Mathematics and Philosophy, edited by Bonnie Gold and Roger Simons

The Random Walks of George Pólya, by Gerald L. Alexanderson

Remarkable Mathematicians, from Euler to von Neumann, by Ioan James

The Search for E.T. Bell, also known as John Taine, by Constance Reid

Shaping Space, edited by Marjorie Senechal and George Fleck

Sherlock Holmes in Babylon and Other Tales of Mathematical History, edited by Marlow Anderson, Victor Katz, and Robin Wilson

Six Sources of Collapse: A Mathematician's Perspective on How Things Can Fall Apart in the Blink of an Eye, by Charles R. Hadlock

Sophie's Diary, Second Edition, by Dora Musielak

Student Research Projects in Calculus, by Marcus Cohen, Arthur Knoebel, Edward D. Gaughan, Douglas S. Kurtz, and David Pengelley

Symmetry, by Hans Walser. Translated from the original German by Peter Hilton, with the assistance of Jean Pedersen.

The Trisectors, by Underwood Dudley

Twenty Years Before the Blackboard, by Michael Stueben with Diane Sandford

Who Gave You the Epsilon? and Other Tales of Mathematical History, edited by Marlow Anderson, Victor Katz, and Robin Wilson

The Words of Mathematics, by Steven Schwartzman

MAA Service Center
P.O. Box 91112
Washington, DC 20090-1112
800-331-1622   FAX: 301-206-9789

Contents

Preface xi

Part 1: Who Are Mathematicians? 1

    Foreword to Who Are Mathematicians? 3

1. Mathematicians and Mathematics 5
   Michael Aschbacher

2. What Are Mathematicians Really Like? Observations of a Spouse 16
   Pamela Aschbacher

3. Mathematics: Art and Science 29
   Michael Atiyah

4. A Mathematician's Survival Guide 31
   Peter G. Casazza

5. We Are Different 48
   Underwood Dudley

6. The Naked Lecturer 60
   T. W. Körner

7. Through a Glass Darkly 71
   Steven G. Krantz

8. What's a Nice Guy Like Me Doing in a Place Like This? 86
   Alan H. Schoenfeld

9. A Mathematician's Eye View 103
   Ian Stewart

10. I am a Mathematician 112
    V. S. Varadarajan

Part II: On Becoming a Mathematician 125

    Foreword to On Becoming a Mathematician 127

11 Mathematics and Teaching 129
    Hyman Bass

12 Who We Are and How We Got That Way? 140
    Jonathan M. Borwein

13 Social Class and Mathematical Values in the USA 156
    Roger Cooke

14 The Badly Taught High School Calculus Lesson and the Mathematical Journey It Led Me To 169
    Keith Devlin

15 The Psychology of Being a Mathematician 183
    Sol Garfunkel

16 Dynamics of Mathematical Groups 192
    Jane Hawkins

17 Mathematics, Art, Civilization 203
    Yuri I. Manin

18 Questions about Mathematics 217
    Harold R. Parks

19 A Woman Mathematician's Journey 227
    Mei-Chi Shaw

Part III: Why I Became a Mathematician 251

    Foreword to Why I Became a Mathematician 253

20 Why I Became a Mathematician: A Personal Account 255
    Harold P. Boas

21 Why I Became a Mathematician? 257
    Aline Bonami

22 Why I am a Mathematician 260
    John P. D'Angelo

23 Why I am a Mathematician 264
    Robert E. Greene

24 Why I am a Mathematician 266
    Jenny Harrison

25 Why I Became a Mathematician 270
    Rodolfo H. Torres

Preface

What This Book is About

The mathematical life is a rewarding, satisfying, and fulfilling one. But it has its trials. As a clinical psychologist has observed, mathematicians and oboe players have a lot in common: they both do something very difficult that other people do not appreciate.

We have all had our fill of the shopworn lament, frequently encountered in social situations, to the effect that "I was never any good at mathematics." Or, perhaps worse, "I was very good at mathematics until we got to the stuff with the letters—like algebra."

Fortunately, there is evidence that the tiresome observations in the last paragraph are not really indicative of the general public feeling about mathematics. Just as an instance, the popularity of the movie *A Beautiful Mind* and the television show *Numb3rs* is an indication of some public fascination with mathematics. The public responded strongly and enthusiastically to Andrew Wiles's solution of Fermat's last problem. There was genuine interest and curiosity about Grisha Perelman's proof of the Poincaré conjecture. Most anyone would like to know more about the shape of the universe. People respect mathematics; they are just intimidated by it.

At the MAA Summer MathFest in 2007, Peter Casazza, Steven Krantz, and Frank Morgan were invited to organize a special session entitled *The Psychology of the Mathematician*. The avowed purpose of that event was to discuss what mathematicians think of themselves and what others think of us. This was a well-attended session, with lively and heartfelt discussions. The speakers went into some detail, and were often quite emotional. As a consequence, MAA Editor Don Albers invited us to produce a volume inspired by our special session. Frank Morgan withdrew from that particular project, and Randi Diane Ruden joined it. So we now have a volume edited by three scholars with diverse interests.

We invited two types of articles for this volume. The plenary articles, generally by well-known mathematical figures, are putatively about the theme of the session: "What Do People Think About Mathematicians?" The secondary articles, rather more brief, are about "Why I Became a Mathematician." Taken together, the two collections of articles paint a varied and multifaceted panorama of ways to think about our profession, our subject, and those who people it.

Among our plenary contributors are Michael Atiyah, Hyman Bass, James Milgram, Steven Krantz, Peter Casazza, Michael Aschbacher, Tom Körner, Tom Apostol, Robert Strichartz, and V. S. Varadarajan. Certainly a distinguished and varied group. Some of the plenary contributors are non-mathematicians who have been close observers of mathematics for many years (such as Pam Aschbacher, the wife of Michael Aschbacher). Several of the plenary writers have devoted a good part of their lives to the teaching and communicating of mathematics. These include Keith Devlin, Sol Garfunkel, Underwood Dudley, and Ian Stewart.

The secondary contributors number among them some notable mathematicians—including Robert E. Greene, Jenny Harrison, John D'Angelo, and Rodolfo Torres. Many of the articles are quite personal. Almost all the authors made a point of telling us how cathartic they found it to write for this volume.

Some of the contributors described above will have their work appear in the second volume of this book. Do be sure to look it up. It will increase your enjoyment and appreciation of this discourse.

The plenary articles are quite broadly distributed. The writers focused on the question at hand in a variety of ways. The results are fascinating, and will be of particular interest to budding mathematicians, budding math teachers, budding math communicators, and in turn their teachers. In order to provide some context for readers, we have divided the plenary articles into two types:

- On Becoming a Mathematician

- Who Are Mathematicians?

These are just rough guidelines, as many articles do not fit squarely into either category. But they will give the reader a hint of what the reading will entail. And they group together like-minded pieces.

What We are About

The job of a computer scientist is to find algorithms that will accomplish certain tasks. The job of an engineer is to make things that work. The job of a mathematician is to develop new theories and establish new ideas and new truths and to teach and communicate them.

And we are obsessive about it. If mathematicians seem to be other-worldly, seem to be hopeless nerds, seem to be excessively compulsive, it is because we are. Trivial worldly matters are of no interest. Nothing measures up to the discovery and establishing of a new mathematical truth, and there is no joy like communicating those truths to our students and colleagues. Nothing has the timelessness, the enduring value, the pure pleasure of mathematical learning and discovery and teaching. This is what we are about.

But one upshot of these considerations is that mathematicians can appear to be isolated. We have trouble communicating with the rest of the world, and the rest of the world has trouble communicating with us. We are perceived to be in an ivory tower, and—God bless us—we may as well stay there. And do no harm to anyone else.

But there is a price to pay for this isolation, and that price is frequently not very pretty. Even in the context of the college or university, we often do not fit in. We frequently are unable to make a good case for our just rewards. We often find ourselves passed over for more trendy or more broadly appealing intellectual pursuits such as genetic engineering or computer visualization or biotechnology. Whereas a geneticist can speak of gene cloning and DNA matching, a computer scientist can speak of bits and bytes and megapixels, a biologist can speak of species verification and evolutionary differentiation, a mathematician has a tough time explaining what we are about. We feel disconnected and unappreciated.

This volume is an effort to reach out. It gives a significant number of mathematicians the opportunity to speak about who they are, where they come from, and what they do. There are also essays by non-mathematicians—ones who know mathematicians intimately—explaining how they see the matter. How do *they* interact with mathematicians and what do they get from that interaction? Does this relationship enrich their lives? What have they learned in the process?

There are many good people who helped to make this book happen. First we thank Jerry Alexanderson and Jim Tattersall for shepherding the book through the Spectrum Series at the MAA. Next we thank Ilya Krishtal and Neal Koblitz for helping with the translation of Yuri Manin's article. Finally we thank Carol Baxter and Bev Ruedi for bringing the book into its final form.

The production of this book has been rewarding for all of us. It has been an opportunity to ruminate and introspect. We have all taken this opportunity to re-live segments of our lives and see what we have learned. We hope that, in the process, we have produced a book that will speak to young people hoping to become mathematicians, math teachers, math communicators, or mathematical scientists. Our view is that this is a glimpse into the personal side of the mathematical equation, one which has been infrequently explored in the past. It should prove to be a productive adventure.

—Peter Casazza, Columbia, Missouri
—Steven G. Krantz, St. Louis, Missouri
—Randi D. Ruden, University City, Missouri

Part I

Who Are Mathematicians?

Foreword to
Who Are Mathematicians?

Certainly one of the main focuses of our special session at the MAA MathFest in 2007 was the question of who mathematicians are and how we think of ourselves. Also we talked about how others (non-mathematicians) think of us. This part of the book is squarely situated in that circle of ideas.

There is no question that many of us think of ourselves in terms of the mathematical work that we do—proving theorems, writing papers, writing monographs, authoring texts, giving talks. But we also think of ourselves as communicators and teachers. Certainly there is no meaning, and little value, to our mathematics if we do not effectively communicate it and teach it to others. And, thanks to Euclid of Alexandria, we have a well-developed paradigm for doing so. For two thousand years we have had logic and the axiomatic method and Occam's razor as the model for mathematical development. And it has served us well.

This mode of thought shapes not only the way that we write but also the way that we teach. For our job as mathematics teachers is to show our students how mathematical discourse works, and *why* it works. We want to teach them many particular ideas and techniques, but we also want to teach them the critical thinking skills that shape our lives.

The essays in this part of the book reflect the value system described in the preceding paragraphs. Several accomplished mathematicians, from many different fields and disciplines, describe their personal take on who mathematicians are and what they are trying to accomplish. Many of these essays are quite personal, and they all serve as role models for young people studying mathematics today—budding math teachers, math communicators, and math researchers.

This part of the book should be assessed and digested alongside the "On Becoming a Mathematician" part of the book; the two taken together form a whole that is greater than the sum of its parts.

— Peter Casazza, Columbia, Missouri
— Steven G. Krantz, St. Louis, Missouri
— Randi D. Ruden, University City, Missouri

I

Mathematicians and Mathematics

Michael Aschbacher

Introduction

I'm going to make a number of observations about the art of mathematics, the people who practice that art, and the ways such people differ from other people. Perhaps enough such observations, here and in other articles in this volume, will add up to some semi-coherent picture of mathematics and mathematicians.

Despite some existing stereotypes, I don't think mathematicians as a class are all that different from the general run of humanity. But it is true that any small group, selected on the basis of the ability to perform some special task, is not unlikely to be a bit different, at least statistically. Here are three differences which I'll discuss later in more detail: Mathematicians learn faster than most people and are better able to cope with novelty. Mathematicians are better able to concentrate deeply for extended periods of time. Some mathematicians have the ability to carry out thought processes at an "intuitive" level, in order to reach conclusions that would be unattainable, or attainable only very slowly, using linear thought processes.

Some of these tendencies, when carried to extremes, can result in behavior associated with the stereotypical absent-minded professor or mad scientist. To the extent that the more random person on the street has any picture of a typical mathematician (and most are totally unaware of mathematics and mathematicians), the picture is probably of a schizophrenic à la *A Beautiful Mind* or *Proof*, or an absent-minded eccentric. Of course it must be admitted that mathematics does have its share of eccentrics, but to my mind that speaks well of the tolerance of the mathematical community and its priorities; the eccentric who produces good mathematics is welcome in the community.

Here are some other things I'll do. I'll describe how students move through the U.S. educational system to become mathematicians. I'll discuss the international community of mathematicians. I'll give my own take on how a mathematician produces mathematics. I hope that this loosely connected collection of small essays will give at least a partial picture of mathematicians and their community.

Mathematics and Mathematicians

To understand mathematicians, it is perhaps useful to begin with a definition or description of mathematics; here is one description with which you may, or may not, agree.

Mathematics is the art of analyzing complex and apparently chaotic situations, finding structure among the chaos, and using that structure to avoid, at least in part, the complexity, in order to establish useful and interesting conclusions.

If one subscribes to this description, then the mathematician is an artist who finds patterns that others overlook; put another way, the mathematician either discovers hidden beauty if you are a Platonist, or creates beauty in unusual circumstances if you are not. But the mathematician is also a craftsman; he or she must take the insights won by intuition and forge them into a useful edifice, using the tools of the mathematician's trade: clarity, precision, abstraction, rigor, ...

The mathematician is also usually a professional. Mathematicians ply their art or craft in various occupations; some solve problems in high tech industries, some create and exploit financial models, some create encryption or decryption schemes for security firms or governments; the list goes on. I will focus on mathematicians who work in academia at research universities, and from now on when I use the term "mathematician" I almost always mean such a person.

Thus, to the outsider, the profession of mathematician is university professor. Indeed almost all mathematicians spend some of their time teaching service courses to undergraduates, some time mentoring graduate students, and some time on service to the department or university; in this last capacity, the mathematician is a middle manager in a certain kind of nonprofit corporation. But when most mathematicians are asked their profession, they will not answer "university professor" but instead will describe themselves as a "mathematician." Mathematicians are only renting part of their time to the university in exchange for a salary and the opportunity to do the thing they like best: mathematics.

A given mathematician may enjoy teaching undergraduates, or mentoring graduate students, or even enjoy committee work. But, almost by definition, to be a mathematician means that your favorite way to spend your time is doing mathematics. Thus we have come to the quality that really distinguishes mathematicians from the rest of the human race: most mathematicians would rather be doing mathematics than anything else.

There are other qualities that, at least statistically, mathematicians are more likely to possess than other people. I've already mentioned two: First, most mathematicians learn faster than other people, and have an easier time learning concepts that are novel. Second, most mathematicians have the ability to focus very deeply for long periods of time. Later we will have cause to consider how much of these abilities the potential mathematician brings to the table, and how much is instilled by the socialization process that turns the student into a mathematician.

Some Background

Before proceeding further, perhaps some background on mathematics and its practice would be in order. Mathematics is in part organized into a number of general subdisciplines, each of which has further specialties and sub-specialties. The major disciplines, from the point of view of mathematical structure, are algebra, analysis, and geometry/topology.

Roughly speaking, algebra consists of the study of structures defined by operations on sets. An operation on a set is just a rule that assigns to each ordered pair of elements from the set a third element of the set. Ordinary addition is the operation that assigns to a pair of numbers the sum of the numbers. To give another example, in a high school algebra course the set consists of polynomials and the operations are addition and multiplication of polynomials. I am primarily an algebraist and my primary specialty is finite group theory. For example, the collection of symmetries of any object form a group, with the group operation consisting of composition of symmetries.

Similarly, analysis studies real and complex valued functions of a real or complex variable. Trigonometry and calculus are early topics in analysis. Finally geometry and topology study mathematical models that formalize our intuitive notions of "space," "distance," "angle," "curvature," etc. A traditional course in plane geometry from Euclid would fall into this category.

But there are also other subdisciplines organized around certain types of problems, rather than around types of structures. For example, number theory studies equations over the integers, and combinatorics is a grab bag containing the mathematics used to solve various problems where there is little apparent structure.

Mathematics is an international enterprise, practiced by thousands of people worldwide. There are formal mathematical organizations or guilds established to support the profession, such as the American Mathematical Society, the Mathematical Association of America, and the London Mathematical Society. But there is also an informal mathematical community, made up in turn of smaller sub-communities, each consisting of people who work in some specialty of mathematics. For example, most of my work has been in finite group theory, so my primary community consists of the finite group theorists. But I'm also a member of the algebraic Lie theory community, and I occasionally consort with representation theorists, combinatorists, homotopy theorists, loop theorists, etc.

The typical professional spends time and interacts with the people at their place of employment, and perhaps also with a slightly larger set of professionals who they encounter in the course of their business. My sense is that most professionals are happy to leave their work behind at the end of the business day, and spend evenings, weekends, and vacations on hobbies or other interests.

Mathematicians interact with the faculty and students at their university, but also have significant interaction with the members of their mathematical sub-community. The sub-community meets in places all over the world at conferences and workshops; also pairs of individuals exchange visits to their respective universities. Such meetings make possible mathematical collaborations, but they also give the mathematician the opportunity to talk to members of that small set of people who speak the mathematician's language and share his or her deepest interest; that language is the vocabulary of the specialty in which the mathematician works, and the interest is in the problems currently of importance to the specialty.

One of the unexpected benefits of being a mathematician is to become a member of such extended communities, and meet friends and colleagues regularly over a period of many decades in interesting places all over the world. So we have yet another characteristic of mathematicians: they are social people who are members of one or more communities of individuals, defined not by geography, but by a shared enthusiasm for a particular branch of mathematics.

There are a number of mathematical institutes devoted to holding conferences in mathematics for a week, or even for several months. Examples include the Mathematical Sciences Research Institute at Berkeley (going by the acronym MSRI, sometimes pronounced "misery"), and the Banff International Research Station in Banff, Canada. But the first of these institutes was the Mathematical Institute at Oberwolfach, in the Black Forest in Germany.

Virtually every week, and year after year, Oberwolfach hosts a conference in some mathematical specialty, beginning on Monday and ending on Friday. The Institute is perched part way up the slope of a large hill, beneath a German forest, and above a small river. Participants live in the Institute for a week, receive three meals and a tea each day between talks, and. in the evening, sit in the Institute lounge, partaking of various refreshments and talking about mathematics or socializing. It is a beautiful setting in which to learn and exchange ideas. The relative isolation minimizes distractions and keeps the focus on mathematics.

Perhaps a word is now in order about the Balkanization of mathematics. As we will see below, in the process of obtaining a Ph.D., almost all mathematicians in the U.S. system take introductory courses in algebra, analysis, and topology, where they learn the basic concepts, notation, and terminology in the corresponding area. Put another way, the young mathematician learns the basic language of mathematics.

But when mathematicians dive into their specialty, they must also learn the special language of that subdiscipline. Because there are so many mathematical specialties, this results in a mathematical tower of Babel, where mathematicians in different specialties can talk to each other at a relatively low level about mathematics, but an outsider to a specialty can converse about the finer points of the specialty with an insider only after a lesson in the language of the specialty.

To mitigate this problem and build a common mathematical culture, mathematics departments at most universities have weekly colloquia, where, in theory, a speaker gives an expository talk about some specialty to a general audience of mathematicians. However, in practice, the goal is rarely completely achieved. Instead the speaker usually overestimates the audience's knowledge of the speaker's field, and the speaker's enthusiasm for his or her problem is impossible to contain, leading to an over-technical presentation which blows away most of the audience. Still, mathematicians are adaptable, and learn to extract nuggets of information from even the most over-ambitious presentations, coming away from colloquia with a bit more insight into specialties other than their own. At worst, one can sit in the back of the lecture room and do one's own work surreptitiously when the the talk is too far over one's head.

Becoming a Mathematician

Let us next spend some time considering how young people come to be mathematicians. At least in the United States, most students are introduced to arithmetic algorithms in grade school, and then move on to manipulation of polynomials (algebra) in middle school and high school. In my day there was a course in plane geometry a la Euclid, which included an introduction to "proof," but my sense is that now such courses are out of style. Those with good quantitative skills may also study some trigonometry or one variable calculus. Usually the fundamental tools of the mathematician—abstraction and rigor—are absent from K-12 "mathematics" courses. Concepts take a back seat to memorization and reproducing definitions and solutions to standard problems.

Thus, until most U.S. students get to college, they are not exposed to what a mathematician regards as mathematics. Their math and science courses may have more clarity and precision than other courses, drawing some students to mathematics. But usually students see nothing of the beauty of the subject during high school.

There are other ways to do things. In the old Soviet Union, mathematicians organized informal "Math Circles" to expose talented young people to real mathematics, and to begin to socialize them as mathematicians. Perhaps in part for this reason, Soviet mathematics was among the best in the world, and produced many talented mathematicians. In recent years, there has been some effort to duplicate such circles in the U.S.

There are various mathematical contests for high school students in the U.S., appealing to the American love of sports-like competitions. In the best of these contests, the problems are deeper than those in the typical high school course, and require more thought and ingenuity to solve. But the problems are still small, and, of necessity, do not involve sophisticated mathematics; so, while such competitions expand the mathematical horizons of participants and teach them problem solving skills, they fall short of giving them a real mathematical experience.

There are also a few summer programs in mathematics for high school students, like the PROMYS program at Boston University and the Ross program at Ohio State. In such programs, students are asked to prove small theorems and are exposed to topics like elementary number theory, which go beyond the usual high school curriculum and involve abstraction and rigor. But only a relatively small number of people participate in such programs.

To reiterate, before entering college, very few young Americans have had contact with what mathematicians think of as mathematics.

What about the college experience? It is my impression that, at most universities, mathematics majors are introduced to mathematics slowly. Bad habits from high school must be overcome. Most students achieve a comfort level with abstraction and rigor only over a period of years. The typical math major eventually takes unambitious introductory courses in abstract algebra and analysis, and possibly a similar course in topology, but there is little or no time for more advanced courses. Few of these students go on to become research mathematicians, opting instead for jobs in high tech industry, finance, etc. Those who do are usually more aggressive, proceed through the program more quickly, and hence are able to take some introductory graduate courses before moving on to graduate school.

On the other hand, at a few selective universities there is a critical mass of stronger students that allows the mathematics department to offer some kind of honors curriculum for its more talented mathematics majors. The introductory courses in algebra and analysis are more substantial, and taught without compromise. The better students take these courses as freshmen or sophomores; they are thrown into the sea of abstraction and rigor and expected to swim. Many adapt instantly and have no problems. Others can cope with the material but take some time learning how to write acceptable proofs. Still others take years to adapt, or never adapt. With a strong foundation established in the first two years, the good students can go on to take graduate courses in the last two years, and often are exposed to some small research experience, where the student works during the summer at a university on a problem supplied by a faculty mentor, sometimes as a member of a small group of students.

The students who do well in such an environment are those who learn and adapt quickly when asked to assume new modes of thought. The topics in a good survey course come rapidly, one after another, and there is the extra complication of having to write up solutions in the style of a mathematician. Focus is necessary; problems are still relatively small, and one knows that solutions almost always require only material from recent lectures; but the problems are also much harder than those encountered in high school, and require relatively long periods of concentration from most students. In short, the students who thrive and are most likely to move on to careers in mathematical research have the ability to learn quickly and focus deeply.

In both types of programs (but particularly in the more competitive programs) there is a Darwinian process in action, selecting the students who can learn rapidly and focus on abstract and relatively difficult problems for a period exceeding the attention span of most students. Students with these abilities are more likely to go on to graduate school in mathematics than their peers who lack them. On the other hand, there is also a socialization process; students learn to think like mathematicians, which facilitates rapid learning, and some acquire the skill of concentrating for extended periods of time. So, are the abilities to learn rapidly and focus deeply ingrained, or instead does a good mathematical education instill them? Probably the answer is: a little of both.

The next stage is graduate school. In most Ph.D. programs in mathematics in the U.S., the first year or two is spent acquiring breadth, in the form of background in a few core areas of mathematics, and then passing some exams to demonstrate minimal competence in those areas. This is a somewhat more accelerated and concentrated version of the undergraduate experience.

Then the student pairs off with a thesis advisor, is given a thesis problem, and begins to acquire the more specialized background necessary to solve the problem and write a thesis. For most students, this is the first contact with research in mathematics, and the first time the student works on a larger, more complex mathematical problem. The student must learn the necessary mathematics, and, more or less at the same time, make progress on the thesis problem. Discipline and focus over an extended period of time are now necessary, as opposed to concentration for short periods devoted to the solution of weekly sets of small problems. Thus at this stage, even larger demands are placed on the young mathematician's ability to learn quickly and focus in depth.

The student also attends his or her advisor's seminar to learn the language and mathematical culture of the advisor's specialty. By the end of the process, at the receipt of the Ph.D. degree, the student has become a member of one or more mathematical communities.

Now comes the first job. For most research mathematicians, this means a postdoctoral position at a research university. By the nature of the situation, there will be at least one faculty member at the new university in the young mathematician's specialty, and perhaps even a small group in which the new Ph.D. becomes a junior member, but at least more senior than the graduate students in the group. Probably there is a seminar in the specialty, further integrating the young mathematician into the corresponding community.

Mathematical Problems

To produce good mathematics, one must work on good problems. A good problem satisfies at least two criteria, both of which depend on the mathematician working on the problem, and on the state of the art in the specialty at that time. First, the mathematician must be able to make progress on the problem, and, second, the results achieved must be of sufficient interest, where interest is measured relative to community norms, but also relative to the mathematician's standards and ambition. Almost all Ph.D. thesis problems are selected by the advisor, since the student rarely has enough knowledge, experience, and taste to find a problem that he or she can solve, and whose solution is sufficiently impressive to secure the student a good first job.

Having left the nest, the new postdoc must begin to select his or her own problems. Sometimes the advisor's research group or the postdoc's new group will have a reservoir of available problems. But eventually, if young mathematicians are to be successful, they must build their own research program.

To illustrate, I'll describe my own experience in transitioning from graduate school to my first few jobs, although my situation was not quite standard, and by now is forty years in the past. My thesis was in combinatorics, but in my last year of graduate school I decided to switch specialties and become a finite group theorist. Thus, for a few years, I had to acquire the necessary background to work in my new specialty and to identify the important problems in finite group theory. During this time I worked on problems that the community had generated and deemed important, and I used some combination of existing strategies and techniques, together with ideas of my own that slightly extended existing approaches. But, after three or four years, I had a good enough feeling for the subject that I could find my own problems, and produce my own strategies and techniques for solving the problems.

Switching Specialties

This also raises the issue of switching fields in mathematics, a practice that does not seem to be very common. However, to find good problems, it is important to be in the right place at the right time, which may require switching fields. If you can identify the right place and learn new mathematics sufficiently rapidly, then you have a chance to move to the right place before the right time disappears. For various reasons, this is easiest to do, as I did, relatively early in one's career. Young mathematicians are still used to spending most of their time learning, while older mathematicians are accustomed to spending most of their free time doing mathematics. Moreover, older mathematicians have more duties and less free time. Finally the established mathematician is a member of a community; to change fields would mean also changing communities, a wrenching experience. It is also easier to move into a field in transition, where everyone is learning at the same time, so the new entrants are less at a disadvantage.

Again I'll illustrate with the example of finite group theory in the sixties and seventies. The building blocks of finite group theory are the finite simple groups, which play a role in group theory similar to the role of primes in arithmetic; that is each finite group is constructed from simple groups, in a suitable sense, just as each integer is a product of primes. When the twentieth century began, the significance of the finite simple groups was appreciated, but after the pioneering work of Sylow, Burnside, and Frobenius, there was little progress in finite group theory for roughly fifty years. Then in the 1950's, local group theory began to take shape in the work of Brauer, Phillip Hall, and Suzuki, while Chevalley began the revolution in the study of groups of Lie type. The order of a finite group is the number of elements in the group. Local group theory studies groups locally, in the sense that it concentrates on "subgroups" of the group in the "neighborhood" of subgroups of prime power order. The finite groups of Lie type are certain finite simple groups analogous to infinite groups of symmetries of certain important geometries; these groups were discovered in the nineteenth century and studied by the Norwegian mathematician Sophus Lie. Chevalley pioneered the use of algebraic methods in Lie theory, as opposed to methods from analysis and geometry.

Then, around 1960, a singularity occurred in the work of John Thompson, who introduced ideas and techniques to the local theory of finite groups that completely changed the subject. By the time I received my degree in 1969, significant new results were appearing almost every day, so that finite group theory was in a state of flux. My lack of background was not as big a handicap as it should have been, since the knowledge base of the subject was changing rapidly. This allowed me to learn local group theory and algebraic Lie theory on the job, along with everyone else, so that I could transition into those communities relatively quickly and painlessly. But to do so, and indeed for more established finite group theorists to continue to function, it was necessary to learn the new mathematics in real time, and adapt to the changing field. This is another example showing that successful mathematicians need to be quick and flexible learners.

The Art and Craft of Mathematics

At this point, I'm going to switch subjects again, and talk a bit about the art and craft of mathematics. I view the process of doing or creating mathematics as proceeding on at least two levels: the formal and the intuitive. When confronted with a problem, the formal level involves establishing formal machinery for analyzing and solving the problem, such as notation, terminology, well-defined concepts, fundamental lemmas and theorems, etc. The intuitive level involves creating "pictures" that somehow encode blocks of formalism, and the manipulation of these blocks in a way which is not rigorous but is often effective. This informal, intuitive thinking allows one to discover facts and lemmas in real time, rather than by much slower formal manipulation of existing machinery.

However in my experience, the intuitive approach is only successful after one has put in some time thinking about the problem to get some insight into the difficulties involved, to develop some familiarity with some representative examples, and to put in place some formal machinery to provide the foundation for an intuitive approach.

Moreover, after spending some number of days analyzing a problem intuitively, and generating some set of conjectures about new machinery and lemmas, it is important to take the time to write out a rigorous treatment of the new material. There are at least two reasons why this is so: first, some ideas and results will not be quite correct; and, second, the process of thinking through these incomplete ideas carefully often leads to more insights, or improvements in existing insights. In short, I tend to alternate between formal and intuitive modes of thought.

To go a bit further, I'd say any good work I've done involves both hard formal work, and intuition. Neither mode of thought can be successful by itself. In a famous article, the well known French mathematician Poincaré described experiencing an insight while stepping onto a bus, which allowed him to solve a problem he'd been working on unsuccessfully for some time. But he didn't think (and I agree) that good ideas appear magically. They emerge at the intuitive level, but only after the groundwork is in place so that one's intuition can feed on enough formal machinery to generate effective "pictures." Sometimes there is the illusion of spontaneity, in the sense that ideas can come, not when one is thinking seriously about a problem, but at a time when part of one's attention is focused on something else. Indeed sometimes when I think I'm very close to completing the treatment of some small problem, but feel frustrated because things don't quite come together, I find it helps to get out of my office and do something else which is not too demanding, like taking my dog for a walk. In the process I'm often able to sort things out, after my mind gets out of a rut, and I'm better able to integrate various parts of the problem.

By the way, I think that many mathematicians often are thinking about one of their problems at a low level, while devoting most of their attention to other business. A different situation is also common: a mathematician may appear to be focused on some routine affairs, while primarily concentrating on mathematics. This can lead to embarrassing moments; for example when your wife is conducting a conversation with you while you are immersed in some problem, and you respond appropriately, with the help of some unused portion of the brain, but later don't remember the conversation. Such incidents reinforce the stereotype of the absent-minded scientist or mathematician.

For a more extreme example, I recall a story (which may be apocryphal) about a well-known mathematician given to such episodes, who drove to a conference in another city with his wife. When the conference was over, the mathematician was concentrating on one of his problems, forgot about his wife, and drove home without her.

In my case, when walking to and from school, I'm usually devoting most of my mind to some mathematical problem. My feet know the route, so I need spend little attention on walking. However for a period when I was younger, I sometimes thought about my problems while driving. Then occasions began to occur where the autopilot in my head would guide the car on a different route than I'd intended. It became clear that I was not devoting enough attention to the driving, so that if something unusual happened while I was in such a state, I might not react appropriately. Since then I don't attempt to think about mathematics while driving, even at a very low level.

Big Problems

I'm going to close with a few comments on big problems and theorems. What is a big problem or theorem? To the graduate student, the theorem in his or her thesis is big if its of sufficient interest to win a good first job. But for purposes of this discussion, a "big problem" is one which is known and appreciated by mathematicians outside of the specialty in which it lives.

A problem could be "big" for various reasons: Perhaps it has a long history and many well known mathematicians have worked on the problem without success, so it has achieved some notoriety. Perhaps the problem transcends its specialty in the sense that it can be stated in the common language of all mathematicians, and almost all mathematicians can understand its significance on the basis of their experience in graduate school. Or, best of all, a solution of the problem may have so many consequences that its importance is evident.

Recall that, in the seventeenth century, the French lawyer and amateur mathematician Pierre de Fermat claimed to have proved that the equation $x^n + y^n = z^n$ has no solutions in positive integers when $n > 2$. The result became known as Fermat's Last Theorem, although the result was not actually proved until 1994, with the last step of the proof supplied by Andrew Wiles, with an assist from Richard Taylor. The Fermat problem is an example of a big problem that even many non-mathematicians can understand and appreciate. While its solution had no important consequences of which I'm aware, it has a very long history, and the many attempts at a solution to the problem led to important advances in number theory.

Probably the most well known and important unsolved problem in mathematics is the verification of the Riemann Hypothesis, which conjectures that the nontrivial zeros of the Riemann zeta function all have real part 1/2. This is a big problem with many consequences (particularly for the study of prime numbers) and a long history.

Here are two examples from my own specialty. At the end of the nineteenth century, Burnside conjectured that all groups of odd order are solvable. All mathematicians learn about groups, and even about solvable groups, in their first course on abstract algebra, so all of us can appreciate Burnside's conjecture. In the early sixties, Feit and Thompson proved Burnside's conjecture, in a classic example of a big theorem. But more important, the theorem, and the techniques introduced to prove the theorem, have had extremely important consequences for the theory of finite groups. Eventually the revolution in the local theory of finite groups, growing out of techniques introduced by Thompson, in part in the odd-order paper, led to the classification of the building blocks of finite group theory, the so-called finite simple groups. This classification is an even more important big theorem, as it has numerous consequences in many branches of mathematics. One of the unusual things about this last theorem is that its proof was not due to one person or small group of people, but was instead an effort in which a large part of the finite group theoretic community participated.

These days big problems are also big in another respect: their proofs are long and complicated. Perhaps in the past good theorems could be proved with short, simple proofs, but by now most of the low hanging fruit has long been picked.

How are big problems solved, and more generally, how do mathematicians approach difficult, complex problems? The solution to a problem that has resisted the efforts of accomplished mathematicians for many years probably requires new insights or points of view. The person introducing the new approach may not solve the problem, but rather set in motion a train of results leading to a solution. Indeed most deep, hard, complex problems with some history are unlikely to be solved only on the basis of one new idea. At best some new idea may make it possible to apply existing machinery in an unexpected way to achieve a solution. In essence, one is just completing the last step in a process involving mathematics developed by the community over some period of time. A more likely scenario is that the community as a whole pounces on the point of view, exploits it, and eventually one or more individuals achieve a solution based on a string of results due to various people.

Returning to our earlier examples of the odd-order theorem and the theorem classifying the finite simple groups, in the first instance Feit and Thompson were able to prove Burnside's conjecture themselves by exploiting character theory recently developed by a number of finite group theorists, and with the use of revolutionary new techniques in local group theory introduced by Thompson. But the proof of the classification theorem required decades of extra work by the entire community, including the development of a much more elaborate local theory of finite groups, plus a complementary theory of finite groups of Lie type.

Finally, this leads to the question of how the necessary new insights are achieved. There are many paths one can take in analyzing any complex problem, so chance must play some role. Perhaps inspiration and illumination amount to stumbling on a new, particularly effective "picture," and in a subject with any depth, effective new pictures are always built from some set of existing pictures, which depend in part on rigorous work.

Still more than mere luck is probably also involved. Hard work almost certainly plays a part, in that it creates the fertile ground nurturing intuition. But the key to most breakthroughs is the ability of some mathematicians to analyze problems in a nonlinear, intuitive manner. This is the reason that computers will not be able to match human beings in solving big complex problems, unless and until they also can think creatively.

2

What Are Mathematicians Really Like? Observations of a Spouse

Pamela Aschbacher

"What's it like living with a mathematician?" I hear this question a lot—probably more than if I were married to almost any other kind of professional. What is implied is: What are "they" really like?—as if they were exotic beings from another planet. And indeed, someone who does math all day long, eagerly thinks about it in his spare time and on vacation, and sees it as "beautiful" is, frankly, NOT like most of us.

I've spent many years living around university research mathematicians as I've tagged along to countless dinners and conferences, spent vacations with mathematician friends, and shared stories with math-spouses over many a walk or cup of tea. Out of curiosity I've informally asked my husband and his colleagues about what they do, how they came to this career, and why they work so hard despite lower pay than in many other positions they could probably hold. I've also discussed with a lot of other math spouses how this career choice impacts their lives. Eventually, it occurred to me that it might be interesting to try to explain to non-mathematicians like my friends what this life is like. In the past year I've also explicitly contacted over a dozen friends who are (or were) also math-spouses to check that my observations and experiences are not anomalous. My attempt here is not meant to be social science research or tabloid fodder. Rather, it merely reflects my own curiosity and desire to open a window on the mathematician's world from my decidedly non-mathematical but nearby vantage point. Unfortunately, my sample is largely limited to the male of the species, as there have been few women research mathematicians of my generation.

By the way, the quick answer to the initial question above is: living with them can be great, but yes, sometimes they do seem to live on another planet.

What Do They <u>Do</u>?

Many of us non-mathematicians have no clue what research mathematicians do when they are "doing math." A good reason for this, they tell me, is that it is *not* what our high school or most college math courses entailed. How were we to know? They *were* called "math" courses. This confounding of arithmetic, algebra, even calculus with "doing real math" undoubtedly is why acquaintances who learn that my spouse is a mathematician often say something like, "Oh, it must be nice to have a husband who is good at balancing the checkbook." Let me just note that this assumption is wrong for many mathematicians. In fact, my non-mathematician friends tend to be much faster at divvying up a restaurant check than the math groups I've eaten with. Mathematicians are good at abstract concepts, not necessarily numbers (unless maybe they are number theorists).

When I've asked mathematicians to explain to me what they do, they usually describe it as solving problems, creating theories, showing that some things are false or true, and/or generating some machinery that constitutes a theory. Some problems are big and important; some have little consequence but are of historical interest. Some modest problems, when embedded in a larger context, build up knowledge to solve the bigger problems. Just to give you an idea of the range of production, a single math proof can range from one page to 1500 pages or more. It might take a few months to several years to prove. Some assertions are around for fifty years or more before someone actually finds a way to prove them.

Part of what makes a "good" mathematician, I'm told, is the ability and willingness to find a problem whose solution would be helpful in the field, to have a productive way of looking at the problem, and to put in the effort to actually solve the problem and write it up so others can use the result. Not so different from other kinds of problem solvers.

What seems to be valued in the math community is having creative, original, new ideas; being able to solve problems and create theories; knowing things that other people don't; and having a large body of work. Unfortunately, like many other creative fields, one-hit wonders are common. Some people solve one good problem but are not able to do much else over the years. As in life so in math: it helps to be lucky—to be in the right place at the right time. People considered to be really good mathematicians have usually done several good things. For example, one not only solved a conjecture that had been unsolved for fifty years, but also developed new techniques to solve such problems. And they seldom have dry spells; with their expert knowledge of what's important in their field, they know what the important problems are and nearly always have something to work on.

As I pondered what it must be like to create theories in math, I have asked several mathematicians: *Is math something like an unseen star, out there to be discovered (the Platonic view), or something we create to make sense of our world (the Kantian view)?* To a person they all told me: "Oh, it's definitely the 'truths' of the world waiting to be discovered!"

I've also asked: *How do you select the problems you work on?* To this I've gotten quite different responses, and when asked in a group setting, the mathematicians were quite surprised at their different approaches, as they had never shared this with each other. One friend told me:

> Problems seem to select me! It's just so exciting. A problem sort of chooses you, and you can't stop thinking about it. At first, you can't figure out how to go about it. You try something, and it doesn't work. You get clobbered! You try something else and get clobbered again! Eventually you get some insights and things begin to come together. Everything is moving. Each day things can look different. It's very exciting. Eventually you've solved it all and it's a great feeling! Then you have to

write it all up (groan). Sometimes I do "pro bono" work, when someone comes to me with a problem, and if I think I can solve it quickly in a matter of days, I'll take it on.

But another mathematician in the same conversation told me he is more active in selecting problems, but he ascribes the difference in approach to the nature of their respective fields of math:

> For me, approaching a problem is different. I don't feel "clobbered." I can usually see lots of different ways into it—it's more a problem of which one to invest my time in working on. I have some sense of what the important fundamental problems are in my discipline and try to spend the majority of my time on them. Also I try to change the thrust of what I work on every few years, partly not to be bored, but also to extend myself in different directions. If I have the necessary math background, I almost never find a problem I can't solve. Sometimes it would take a lot of time to learn enough background to be able to solve a problem so I have to decide whether it's worth it.
>
> I think our differences reflect the differences in our fields of expertise. In my field there are plenty of techniques for approaching a problem so there's room to create your own techniques. There's a lot there to try to understand, and it's not a problem to try to find a foothold, as in his field.

I wonder if anyone has studied this issue? Do problem-solving approaches vary more within fields or between fields? If approaches tend to be idiosyncratic, they may reflect some hard wiring in the brain we have yet to understand. Perhaps functional MRI will eventually be used to explore questions like this.

What is This Passion for Math and How Does It Develop?

The mathematicians I've known see beauty, wonder, harmony in the universe through mathematics and have a very aesthetic sense of it—something most non-mathematicians can't quite understand. We might feel satisfied when we solve a problem but pure math is the intoxicating discovery of something previously unknown, using very abstract thought. To me it seems a bit like discovering the existence of a galaxy that we can't actually see but must infer from its effects on other bodies in the universe. It is fortunate that mathematicians tend to be motivated by their passion for knowledge creation, not for fame or fortune, since academic salaries are not high, given the time and effort devoted to advanced degrees and postdocs, and the average person could probably not name a famous mathematician (beyond perhaps Pythagoras or Euclid). Within the field, however, people can gain some small fame from having their names assigned to the things they discover or prove, e.g., Lie groups, Frobenius groups, Suzuki groups. Still, according to the people I've known, it's more about the discovery itself than being famous as the one who made the discovery.

Most of the mathematicians to whom I've posed a question about how they were drawn into math have said they feel they are just born that way—and that there is a strong self-selection process in which many people are turned off and away from math but a few are drawn towards it by its beauty and elegance—words that non-mathematicians typically do not use about the subject!

However, as one math spouse explained it, while her husband's love of math seemed innate, his children may well have benefitted both from the genes he passed on and the model he provided:

> When he was young, he fell in love: (1) with painting and art and (2) with math. He says that for a while the two were tied but painting was never in the lead. Eventually he realized that he was much better at math. If I, after knowing him for some 35 years, had to analyze his choice, I would say that his passion for the beauty of math was what overtook all else. I think for him it's almost an addiction to the art of beautiful mathematics: its form, its balance and its ability to clearly explain the almost unexplainable. Whenever I've asked him to try to explain his work to me, his eyes sparkle and he gets an animation in his voice and gestures that are not present at any other time. I believe that my husband's love of mathematics is so heartfelt that no one ever needed to explain that math was beautiful to our daughters. They truly felt that, like the art we love and have in our home, math was an art to be appreciated and that the pursuit of trying to understand it was a very worthwhile undertaking.

A few mathematicians told me they found their passion for math early in life. For example, one told me that like many young boys, about age 9 he was very interested in rockets, so he got a book about them from the library (this was pre-internet days of course), but he could not understand it. He showed it to his economist father, who said, "No wonder, it's full of math," and proceeded to teach his son some math. The son "got caught up in the wonder and beauty of it" and kind of lost interest in the rockets.

A woman mathematician told me that her interest in math grew out of a high school interest in science:

> My exposure to astronomy at the Green Bank Observatory in high school was a turning point for me. I was fascinated by all the machinery, by the vast reaches of space, and the fact that we could see objects so far away. I started college as an astronomy major, taking math and physics, and discovered that I really loved the physics and could study astronomy through physics without going to remote observatories. And then I discovered that *math* could do everything, and that I really loved it.

In fact, most of the mathematicians I've discussed the topic with did not discover the beauty of math until college or graduate school, and it was sometimes by accident. One told me that he became interested in math when he couldn't get the history and philosophy courses he was interested in at a big public university and happened to take a decent math course instead. Many said they entered college more interested in other subjects such as history, literature, engineering, or physics, but they discovered that they liked math and did better in those courses. For example, one told me how a combination of college experiences ignited his math interest:

> In high school I didn't think a lot about what I wanted to be, I just sort of drifted through life... When I went to Caltech, I took all the usual courses. I didn't like the

chem lab—hated the details—didn't do so well in physics class with Feynman. He gave problems you had to solve by guessing the right math model. I was frustrated because the TA liked my model one week, but the next week after he'd learned what the professor wanted, he said my answer was wrong. I did better on math-related parts of physics than the other parts and I had some interesting math classes in analysis and algebra where the professors were quite enthusiastic. So sort of by a process of elimination, I became a mathematician. I liked it, and it seemed to suit my interests and the way my brain works.

Sometimes a college instructor gave pivotal advice, as one spouse told me:

My husband did not think of being a mathematician in high school. He was under family pressure for a particular career, and that's the direction for which he prepared himself, not thinking of anything else. Fortunately in college, however, his professors recognized his math abilities and urged him to think about that field instead.

For some, a career as a research mathematician is fraught with practical challenges—job offers may not be where the spouse can get a job or wants to live, academic postdocs and starting salaries are lean, and at times the job market has been so tight that people left the field in frustration—but math seems to have a strong pull for many of them nonetheless. One woman mathematician I know had several good academic positions but gave them up to live in the same country as her husband and children, where the only relevant position she could get for several years was teaching in a technical college. I saw her at a math meeting after several years' absence, and she explained she really missed being a research mathematician and had come to assess whether it seemed feasible to reenter the field. Fortunately she did and eventually was able to secure a research position in her adopted country. Another friend left math a few years out of graduate school when the job market was very tight and he could not get a tenured position. He became a physician and enjoys his work, but after a while he missed the intellectual challenge of math. Fortunately, he found a way to do some math on the side with scientists at the local state university.

Doing math is just so compelling for them that most mathematicians I know would rather be doing math than any other job in the world. Some of them love teaching classes too, and most agree out of duty to be a department chair, a dean, or an editor at some point, but doing math remains their deep calling:

As Chair of the math department, my time is filled with administrative tasks and I don't have enough time to work on my math. But in two years I'll be able to give up that job and give up editing a journal so I can devote myself to my own problems for two years—ahhhh, heaven!

So Do They Really Write Theorems on the Backs of Napkins?

In a word, yes. One thing I learned early on was not to throw away scraps of paper with math symbols on them, no matter how old or grungy. The best I could do was to create a special place to corral the odd bits so they would not go missing. I've asked my husband about this and he says it occasionally helps to write something down or draw pictures or graphic representations of his thoughts. What are these? Little circles and squares to help him remember different mathematical objects and their relationships (like "something is the product of two groups with a third group acting on top of them"—whatever that means), and he uses both standard and unique elements in his own personal shorthand. He jots them down on napkins, receipts, the backs of other pages, anything at hand when the thoughts come.

Concentration and Parallel Processing

So what are mathematicians like when they are working? Well, in what other job can you work while lying in bed, waiting in line at the DMV, or singing along with a CD? Some mathematicians really can do these things while they are deep in mathematical thought. I suspect that, for many of them, math is nearly always going on somewhere in their brains. When I asked my husband what it takes to be able to do math *well*, he gave first priority to their amazing powers of concentration.

> You have to be able to focus and concentrate for an extended period of time and follow a train of thought without writing it down. It's an absolute necessity in order to be a mathematician. But it can be dangerous. You're shutting out everything else. For example, you can't possibly allow yourself to do it while driving!

He often thinks about math when falling asleep or in the early morning as he's beginning to waken but it can also prevent him from relaxing enough to sleep. He thinks a little about math while walking his usual route to school and likes having that time to do this. Somehow he is able to parallel process enough (perhaps using a kind a of auto-pilot) not to stumble, and he says when he has to cross a busy street on the way to campus he changes focus to "reality."

To me, this ability to concentrate is one of the most extraordinary things about mathematicians! How can they manipulate abstract concepts like "groups" in their minds without writing much down and despite the sort of distractions that irritate the heck out of the rest of us? My husband claims that with practice over the years he's gotten better at concentrating. This kind of concentration is just beyond anyone else I know. Most of us math-wives are in awe of it—although at times it evokes frustration and/or laughter.

At a wonderful math meeting we attended on Crete, about thirty mathematicians and assorted spouses had a free afternoon and went for a guided hike in the mountains (populated mainly by sheep, a shepherd or two, perhaps a few wolves, and one or more caves where Zeus was purportedly born). After the hike we were to have dinner at a tiny village at the bottom of the mountain, then a bus would return us to our lodgings in another tiny village on another mountain. (Did I mention it felt like we were 1000 years back in time, out in the middle of nowhere, and could not speak the language?) About 5:00 pm, when it was time to head down the mountain to dinner, two mathematicians were nowhere to be found. We spent an hour scouring the area for them to no avail and began to get very worried about them. Eventually it got so late that the hike leader continued the search herself while the rest of the group virtually ran down the mountain to beat the quickly descending darkness (of course we hadn't brought flashlights since we hadn't expected to be out so late). With speed imperative, most of the group began to scramble down a steep and fairly straight downward animal track in a crevasse, while a few others took a steeply winding trail covered in chewed up concrete (called a "road"). About an hour after darkness fell, the groups had made it to the bottom, although one person had badly sprained an ankle navigating the crevasse. In the village we discovered the two missing persons, completely unaware of our troubles. They had been deeply discussing math since the afternoon and, oblivious to our potential concern, just ambled away from the group and headed down the road, merrily talking math all the way.

And then there was the fellow (who will remain nameless) who, deeply involved in thinking about math, left a math meeting at its conclusion without remembering that his wife had come with him on that trip!

Another mathematician was at home with his son while his wife was at work.

Of course he was doing maths, leaving our son, who was not yet two, to amuse himself. After a while he was disturbed by a knock on the door. It was the village policeman (this was in the far-off days when villages had policemen). "I just thought you'd like to know that I've spent the last half hour following your toddler around the village," he said. "I didn't want to talk to him in case it frightened him, but he's been wandering the streets happily and has just come home again."

To be fair, the husband had no idea that his son knew how to open the back door—but, according to his wife, he had been totally oblivious to his son's absence.

A friend told me that her mathematician husband had been so focused and deep in mathematical thought that he never even noticed a moderate size earthquake. For him, being able to concentrate to this degree is just part of his nature—like breathing or sleeping. Indeed, at times, when working on a most difficult problem, he even works in his sleep. Now and then, he will wake up having solved a large equation that was an integral part of the "big picture." My friend does most of the driving as she worries that her husband could easily fall into thinking about math and miss his turn or get into an accident. My own husband has adopted a way of strategically driving slightly ahead of the flow of freeway traffic specifically to avoid slipping into "mathland." If he is concentrating hard enough on driving, his mind doesn't wander to math.

Long before computers could do parallel processing, married mathematicians had perfected this ability. Many of them have adapted to the needs of their partners by appearing to listen while actually deep in thought. A wonderful example of this skill occurred three days after two of our friends brought their second son home from the hospital. It was a Monday morning and the wife had to drop the older son off at pre-school. Before she left, she walked over to where her husband was preparing his classes, reminded him that the baby was sleeping upstairs and told him that he should wait until she returned in a few minutes before leaving for classes. He looked at her, nodded and said, "See you soon." She returned fifteen minutes later to find the baby still asleep upstairs and her husband gone. She called him at school and asked, "Did you forget something?" "What do you mean," he replied? "The baby," she said!

I learned early on in our marriage to spot "that look in his eye" that reveals he is thinking about math while carrying on a conversation with me. His responses are usually quite appropriate in the conversation, but there is a slight delay like a phone call between the US and Europe. It feels as if he is working about twelve feet under water and then comes up to a depth of about two feet when I ask a question—but he's still under the water. I eventually learned to politely ask, "Are you thinking about math?" The reply is usually, "... mmm, a little."

A friend told me her mathematician husband has always had the most amazing ability to concentrate totally on a problem, but at the same time his mind is filing away what she is saying to him.

> I will tell him something that I need him to remember, see that he is abstracted, and say, "You didn't hear a word I just said!" He will slowly (really just a few seconds) redirect his attention to me and repeat back what I just said. We developed one coping mechanism for this in our early years, when we were driving across France after a conference. He was driving and I said something that I felt required at least an acknowledgement. I stopped chattering and waited. Nothing. I said, "T.R.A.A." He looked startled and asked what I had just said, so I repeated, "T.R.A.A. —That Requires An Answer." And he answered the question within seconds. For a couple of years I used "T.R.A.A." whenever I needed his attention.

Many of us mothers have learned to multi-task, for example, to talk on the phone while cooking dinner and answering children's questions at the same time, but it's not the same. None of those tasks requires the same abstract depth as math. One of my friends noted: "At work, I could talk on the phone with a student while answering email and thinking about a report I was writing—but none of those requires the same focus as mathematics!"

I have been interested in this issue of concentration since first encountering it many years ago and have discussed it from time to time. For example, my husband likes to listen to music while doing math, but a friend who's a mathematician says he just can't do that; the music gets in the way. But not only does my husband listen to the music while thinking math thoughts, he sings along with it. It got to be a joke when my mother would call, hear him in the background from the other room, and casually ask, "Is that [him] singing?" and I would say, "Yes, he's doing his math." I also finally discovered why he watches B-movies on the weekend. For years I just figured he had lousy taste in movies, but I noticed he didn't seem that involved in them and finally asked why he watched them. It turns out that they provide just the right amount of white noise for him to be more productive in math.

> Sometimes I can think better when part of my brain is sort of distracted by music or something on TV that I'm half-watching. There's sort of less tension. If it happens, it happens; if it doesn't, it doesn't. Sometimes it's better not to try too hard, not to think in too linear a fashion.

He also says he usually sits at the back of the room for talks at math meetings so that if the talk is boring or out of his field, he can think about his own math. He says his colleagues seem to adopt the same strategy so no one is offended by this practice.

I suppose mathematicians aren't necessarily different from others who work largely with their minds. Theoretical physicists are probably quite similar; professional chess players are famous for their memories. But face it, most of us non-mathematicians just can't do this with our minds.

Math Identity—How Do They See Themselves?

I imagine the other authors will say far more about this than I could. However, I'll make one observation that they might miss because I believe they take it for granted. I've noticed that, when asked in a mixed group what they do for a living, they do not call themselves a "college professor" or "math professor," which would be more broadly understood. Rather, they refer to themselves as "mathematicians," even though the listener might not know what that really means or might be put off by it (usually saying something like, "I always hated math in school," or "I never was any good at math.") I notice the same happens when my husband has to fill out tax forms or other paperwork asking for his profession. Even at the risk of being misunderstood or put down, they see themselves as mathematicians first—i.e., people who discover/prove mathematical truths. This is the essence of who they want to be, the part of their job with which they most identify, the portion that carries the most importance for them, even those who also enjoy and may excel at being a teacher, department chair, or dean (titles to which some listeners might accord more status than to "mathematician").

I believe their strong sense of identity is nurtured and enhanced by working within the same small community of research mathematicians all their work lives. Over time, they accrue shared experiences (like the Crete hike), compare common concerns (like how to recruit the best graduate students and young faculty), and develop fluency in their common math language. Since much of their interaction is at conferences and meetings where the point is to publicly share and critique their work, they get to know each other well, assessing each others' professional strengths and weaknesses. They quickly learn whose talks are worth going to, whom to ask if they want information about a topic. They also get to know one another's personal characteristics revealed over the many hours spent together at professional meetings, having meals or drinks together, discussing university and world events, and participating in recreational events like walks or tours planned for meeting participants and their families. There is a wonderful comfort in working with people—a kind of math family—whom you know well and can trust to share your work ethic and professional values, to appreciate your enthusiasms. No wonder they enjoy their professional meetings so much. They provide warm reward for the long hours and hard work that doing math inevitably entails.

At the same time, mathematicians' sense of identity is probably also reinforced by the fact that many non-mathematicians do not know what they do and tend to hold them at arm's length, ascribing to them various negative stereotypes. I perceive that many mathematicians are saddened when others proudly proclaim about themselves that they are "not good at math," implying that this is a positive attribute and that they have no desire to appreciate mathematics. This seems to be an advanced version of the anti-intellectualism rampant in high school and American society in general. Mathematicians are smart, so they are doubly rejected, but from that rejection is born a stronger group identity. In addition, the math group is, overall, a really nice, interesting group of people and one I'm glad to be part of!

What Do They Value?

Mathematicians are committed to ideas, to learning, to developing knowledge for its own sake. The usual fame and fortune seem not particularly seductive for them. Many math spouses told me that this quality is one of the things they love about them. We respect their dedication to their work and the high standards they have for it. Most mathematicians are excited by the challenge of their discoveries, and they tend to feel lucky that someone will pay them to do what they love. There are a few math prizes for extraordinary achievement, some with a little money attached, but it's not usually a lot and is basically for travel to work with others. They can be competitive, but the stakes are lower than in other fields, so they can afford to share their unfinished work, unlike scientists in many fields. Few of their math colleagues are interested in or capable of beating them to a solution. Since their work is public and its value so readily apparent, they cannot fool their colleagues. Their primary reward then is the hard-earned recognition and respect from their peers.

Social Skills

The stereotype is that mathematicians aren't very social, but in reality there is great variability in their verbal and social skills. Some of them can't be bothered with small talk, but others are great schmoozers who can tell delightfully entertaining stories. Many may seem slightly introverted or shy at first, but once you get to know them they are quite personable. As one math spouse noted, many mathematicians like her husband are interested in a lot of different subjects (e.g., food, wine, art, travel, science, languages, music, and theater) so it is easy to find a common interest when meeting someone new. Some of them thoroughly enjoy their roles as teacher, advisor, department chair or dean, which all require a lot of social interaction. Some are like a parental figure for their students, maintaining fond contact with them even after graduation.

How Has Being Married to a Mathematician Affected Our Family Life?

In thinking about this question, I have concluded that the *social* and *geographic* aspects of mathematicians' work can strongly influence their families' lives in positive ways. Working in a U.S. university has an international flavor for mathematicians, as many of their European, Asian, Indian, and former Soviet Union colleagues spend years of their careers here. In addition, research mathematicians often attend several math conferences or meetings each year with others in their field, at which they discuss their latest work and, in the process, get to know one another. These meetings are quite international. When held in the U.S., there are still many foreign mathematicians who attend as visitors or are grad students, post docs, or instructors temporarily studying or working in the U.S. For meetings held in Europe, probably half the attendees are from the U.S.

Math meetings are relatively easy for mathematicians (and often their families) to attend. Since mathematicians don't have to run a lab like many other scientists, it is relatively easy for them to get away for such meetings, and their travel expenses tend to be paid by grants from the U.S. government or a foreign one. If the meetings are not during the summer, mathematicians just have to find someone to teach their classes and avoid leaving during times of important committee work (e.g., during admissions season). Some can even arrange to do all their teaching in two terms, leaving the third free for travel. English is nearly always the language of the meetings regardless of where they are held.

The majority of meetings my husband attends are focused on his particular math specialty, and his colleagues around the world number perhaps only 200–300 people, about 50–75 of whom might attend a typical meeting. When feasible, spouses and children may accompany them. Through repeated shared experiences with this relatively small but international community, mathematicians and their families knit together friendships all over the globe. When I accompany my husband to a meeting, I often have the chance to see other math spouses who have become friends over the years. Even if much time has elapsed since we last met, our common experiences and interests allow us to pick up nearly where we left off years ago.

Sometimes families can live in one place for a month to a year, living more like locals than tourists. Their children can go to the neighborhood school and acquire a foreign language. For the month we lived in a Paris suburb when our daughter was 3 $\frac{1}{2}$, she began to babble nonsense syllables with a French accent when playing with other children in the communal sandbox. Three years later when we were in Oxford for three months, she developed a temporary English accent. This early language experience may well have influenced her later interest in and ear for languages.

I realized how *much* I cherish these opportunities when I chatted with some tourists while we waited in line in the Piazza San Marco to see the duomo in Venice last year. They had taken a behemoth cruise ship to visit five big European cities in as many days, and they loved being able to sleep and eat on the ship. I, on the other hand, had just spent two weeks in Venice while my husband taught a math course to E.U. grad students. I must have walked nearly every street in Venice, talking to local artists and students, eating cichetti for lunch and fabulously fresh seafood for dinner at little places recommended by our local friends. I found tiny museums and fascinating local shops off the tourist track, watched a traditional boat race from the venerable Grand Canal home of a local mathematician, and attended a math soiree on the small island where Venice originated. Each day was a new gift of delicious experiences. How could I not feel far more fortunate than those tourists!

All this exposure to the people, language, culture, and food of so many other countries helps us look outward, become interested in others, and consider different ways of being and thinking than we otherwise might. This is not unique to mathematicians, of course, but they may be able to do more of this sort of travel than many other professionals, whose meetings may be larger and less frequent, usually located within the U.S., or tend to resemble typical tourist-style travel. I feel incredibly grateful to have visited so many fascinating places. The many opportunities our family has had to develop friendships with people all over the globe have greatly enriched our lives—largely due to my spouse being a research mathematician.

Language and Communication in a Math Family

I've found that the language of our family conversations can be slightly different from non-mathematician families, particularly an emphasis on precision of language and thought, which is crucial in math. My husband is very succinct and logical (another priority from the math world) while I use a much longer, less direct narrative structure. (One mathematician may have taken this to an extreme. When asked why he preferred the female mathematician he married after divorcing his non-mathematician wife of 25 years, he said, "I don't have to talk to her. We can communicate mathematically.") My husband is also excellent at calling out the talking heads on television who use language very loosely with ill-defined terms, imprecise language, and sloppy thinking. In addition, the vocabulary of math, logic, and probability infiltrates the everyday speech of our family and others I've known. When we use these terms outside the family or math community, however, it can result in odd sideways glances. Here's a sample with some translation:

Math terminology	Normal language
John is the random variable.	As soon as we know what he wants to do for dinner, we can plan the rest of the day.
It's a non-trivial task.	It's hard.
It's hopelessly difficult.	It's really hard.
It's intuitively obvious.	Any dolt should know this.
They have to worry about the 2-body problem.	It's hard for a married couple to both get academic jobs in the same location.
Boy, the chance of that happening is epsilon!	A VERY tiny chance, close to zero.
We thought it was impossible to be both sexy and a mathematician, but John is the existence proof.	John's being both proves it's not impossible.
Jane takes her interest in photography to the nth degree.	Her interest is a bit extreme (she just bought a $2000 camera).
We are asymptotically approaching an agreement on which movie to see.	We won't be seeing a movie tonight.
Their friendship is asymmetrical	John gets to do what he wants a lot more often than Jane does.

Math also influences the way my husband remembers things. For example, once he revealed that he remembers my birthday by recalling that it's a prime number.

Math Spouses

I have not noticed obvious patterns in the types of women who become math spouses beyond being smart, well-educated, self-sufficient, and accomplished. They tend to have careers outside the home, but these span a very broad range of fields such as psychology, social work, science research, history, educational research, higher ed administration, law, business management, visual arts, music, nursing, teaching, and writing. A precious few are mathematicians themselves. For the most part, we non-math spouses do not substantively understand the math that our husbands do, but fortunately we don't need to. We quickly learn the structure and regularities of the job, so we can be sympathetic about whether their department will hire the person they want to recruit or whether they will have to put much of their own research on the back burner for three years if they agree to become the next department chair. In the more successful marriages, the mathematician will do the same for his spouse! In the early years of our marriage I was a bit awed by the depth of his focus on math, compared to my multiple but shallower interests both within and outside my career field. Eventually I realized that we each have a different kind of intelligence and a unique mix of talents, and that both are valuable. Over the years, if we are lucky or willing to work on it, we develop mutual respect for each other's strengths as well as tolerance and support for each other's weaknesses. One of my friends captured this spirit well when she described their relationship.

> At first, I admit, there was an adjustment. I had thought of myself as a fairly bright person. It was eye-opening to first understand how fast his mind works, how incredible his memory is and how extensive his abilities are. From having memorized all of Hamlet at one point, to memorizing the first book and some of the second book of the *Iliad* just for fun, to being able to sing Brahms's *Four Serious Songs*, Gilbert and Sullivan, and Tom Lehrer. But I came to realize that his world still needed my world, and my world needed his world. After all these years, I'm still awed, frustrated, uplifted, inspired, and fascinated by him.

On the day after I sent a draft of this chapter to a math-spouse friend, she sent me this return email, which seems to confirm these observations:

> Hi, Pam,
>
> Coincidence of the century. This afternoon we were at a talk by a very prestigious university prize winner. The title of his talk was, "I Want to be an Algebraic Geometer" and he talked briefly about why he became a mathematician and why algebraic geometry. His reasons were "the beauty of math" and the "joy of algebraic geometry." At the end of the talk he thanked his wife of 28 years, who had given him the peace to do his research.
>
> Love, J

3

Mathematics: Art and Science

Michael Atiyah

"Cogito ergo sum" said the mathematician and philosopher René Descartes, highlighting the fact that mathematicians are thinkers. Our brain is our only tool and we live and work in a world of ideas. In theory we could work on a desert island or, like Jean Leray, in a prisoner-of-war camp. All we need is peace and solitude to allow our brains to explore the mysterious world of mathematical ideas.

This self-sufficiency is no doubt one of the reasons why the general public view the mathematician with a mixture of awe and incomprehension. It also explains why, in extreme cases, mathematicians live at times on the verge of insanity. Total immersion in the abstract world of ideas is a demanding discipline and when taken to extremes it can overwhelm.

But mathematicians are not unique in this respect. Art can be as harsh and many famous artists have found themselves on the thin line that separates creative concentration from delusion or depression as with Goya or van Gogh. Isolation and introversion can foster the creative imagination but they can also be psychologically disturbing, as with the famous Austrian logician Kurt Gödel.

But if mathematics shares the freedom of creative thought with art, it is also close to natural science and so gets anchored to the physical world. Mathematics faces both ways; it is both art and science, and its practitioners serve two masters, Beauty and Truth. How to reconcile these apparently contradictory gods is a conundrum that lies at the heart of mathematics, and each mathematician has to solve it in his own way.

That beauty and truth can, on occasion, live happily in harness is clearly shown by Newton's deduction of the elliptical orbits of the planets from the inverse square law of gravitational attraction. This is a beautiful explanation of an important scientific fact.

Both art and science are human activities that have their own evolution, but there is a fundamental difference. Science is hierarchical, all discoveries build on earlier work and aim at a unification of knowledge. Newton would recognize the advances made by James Clerk Maxwell and Einstein. It is much less clear what Bach would have made of twentieth century music.

A related aspect of the difference between art and science is that, while both depend ultimately on the spark of individual genius, science is collaborative in a way that art is not. Mathematicians frequently work together in a very detailed way and many papers have joint authors. Musicians may admire the work of others as with Haydn and Mozart but they do not produce joint symphonies. Painters and sculptors may employ assistants, but they claim the final product as their own. A joint painting by Michelangelo and Leonardo would have been regarded as an oddity.

By contrast mathematical theorems frequently have double-barrelled names, and with equal partners, such a creation can have more than one parent.

Mathematicians, like painters or musicians, come in all shapes and sizes. Some are solitary, some are gregarious, some look the part while others look like the common man. What they have in common is a passion for mathematics that drives them to pursue the ultimate goal of understanding. We are, as Descartes noted, all thinkers, but we may think in different ways and work in different areas. We have to obey the rules of rational thought and search for rigorous proofs, but the source of our inspiration lies buried more deeply in the imagination. Even Karl Weierstrass, the epitome of rigor, said that a creative mathematician had to have the soul of a poet.

4
A Mathematician's Survival Guide

Peter G. Casazza

1 An Algebra Teacher I could Understand

Emmy award-winning journalist and bestselling author Cokie Roberts once said:

As long as algebra is taught in school, there will be prayer in school.

1.1 An Object of Pride

A mathematician's relationship with the general public most closely resembles bipolar disorder—at the same time they admire us and hate us. Almost everyone has had at least one bad experience with mathematics during some part of their education. Get into any taxi and tell the driver you are a mathematician and the response is predictable. First, there is silence while the driver relives his greatest nightmare—taking algebra. Next, you will hear the immortal words: "I was never any good at mathematics." My response is: "I was never any good at being a taxi driver so I went into mathematics." You can learn a lot from taxi drivers if you just don't tell them you are a mathematician. Why get started on the wrong foot?

The mathematician David Mumford put it this way:

> I am accustomed, as a professional mathematician, to living in a sort of vacuum, surrounded by people who declare with an odd sort of pride that they are mathematically illiterate

1.2 A Balancing Act

The other most common response we get from the public is: "I can't even balance my checkbook." This reflects the fact that the public thinks that mathematics is basically just

---

The author is supported by the National Science Foundation—Division of Mathematical Sciences, the Air Force Office of Scientific Research, and The Defense Threat Reduction Agency.
E-mail address: pete@math.missouri.edu

adding numbers. They have no idea what we really do. Because of the textbooks they studied, they think that all needed mathematics has already been discovered. They think "research" in mathematics is *library research*. They have no idea that mathematicians can't balance their checkbooks either—although for reasons different from theirs.

1.3 Accounting to the Public

The public sees us as slightly mad geniuses since we take for granted things they cannot even imagine. They see us as aliens who are just visiting this planet long enough to make their lives miserable. They are not sure if they should be talking to us or running for the exit. They are pleasantly surprised if they discover we can hold a normal conversation with a mere mortal. We like appearing as "geniuses" to the public. If we must have a *false mystique*, this is probably the best we could ever hope for. The problem is that their definition of "genius" is quite different from ours. That is why they can think of actors as *creative* and mathematicians as *accountants* who can balance checkbooks.

1.4 We Have Major Problems

This view of mathematicians as "geniuses" creates problems. Many students are discouraged from entering mathematics since their teachers don't see them as brilliant enough to be in the field. Believing that success comes from innate talent takes away any control you have over your career. And after being classified as a genius, you try to live up to this expectation only to discover that brilliant inspirations are few and far between. And, if you manage to do something spectacular, you can easily become obsessed with trying to outdo this by working only on the *major problems* which have ruined the careers of many fine mathematicians before you.

1.5 Our "15 Minutes of Fame"

We keep waiting for the public to give us some positive form of recognition. But, even when they defer to mathematics, it is often facetious as exemplified by the common expression: "You do the math." And it does not help with the public to remove our mantle of *mathematician* and replace it with *faculty* since it just lands us on the sword of journalist William F. Buckley:

> *I would rather entrust the government of the United States to the first 400 people listed in the Boston telephone directory than to the faculty of Harvard University.*

Perhaps our relationship with the public is best summarized by the following scene from the television series *Law and Order*. Two police officers are standing over a dead body in a high school classroom.

> First officer: *An art teacher. I can't believe she ever hurt anyone.*
> Second officer: *An algebra teacher I could understand.*

2 "All My Imaginary Friends Like Me": Nikolas Bourbaki

The famous satirist and mathematician Tom Lehrer (Isn't "Lehrer" the German word for "teacher"?) once said:

Some of you may have met mathematicians and wondered how they got that way.

2.1 Fulfilling Careers in Mathematics

Mathematicians form a broad spectrum of personalities from normal to isolated, introverted to outgoing, etc. We have a different definition of *normal* precisely so that we can declare ourselves to be in this category. Mathematicians from a very tender age may see themselves as different. Worse, those around you may see you as different. The word "nerd" arose so people would have at least some definable category to put us in. This partly comes with the territory. The very traits that make us good at mathematics work against us in society. A minimal requirement in mathematics is a certain level of being *obsessive-compulsive*. An obsession is a "persistent recurring thought," while a compulsion is "an action a person feels compelled to carry out over and over." What abnormal psychology texts see as a "disorder," we embrace. They refer to the bad side effects of *obsessive-compulsive behavior* as:

"Emphasis on logic and reasoning over feeling and intuition."
"Keeping everything in order and under strict control."

O.K. But I am still waiting for the *bad* side effects.

As if this isn't enough, take a look at *Asperger's Syndrome*. "A condition on the autistic spectrum. It includes repetitive behavior patterns and impairment in social interaction." Finally, they end with the punch line:

These characteristics can often lead to fulfilling careers in mathematics, engineering and the sciences.

Thank you. We needed that recognition. The main point is that the very traits that make us good as mathematicians make us not so good at social interactions. So many mathematicians are quite introverted. Luckily, in a group of mathematicians, you can easily tell the extrovert. She is the one looking at *your* shoes when she is talking to you.

2.2 Adapting to Intelligence

After reading about these "disorders," I am not sure if I should be removing my two favorite signs from my wall:

Gone crazy. Be back shortly.
Anything worth doing is worth overdoing.

People like to say there is a thin line between genius and psychosis. And there are many famous cases where mathematicians fell over the line. (Go see the movie *A Beautiful Mind*. And, to get a clearer picture of how Hollywood sees us, make sure you also see: *Proof*,

Pi, and *Good Will Hunting*. But remember that this is *fiction* and represents just how others see us.) The problem is that mathematics did not make us "legally" insane, or we could walk away and take care of it. Rather, it was precisely these characteristics that drove us into an area that finds all our strange behaviors completely normal—even desirable.

Employers like us because we question everything—even those things they have held sacred forever. This gives them a chance of making real needed changes in their companies. But these same qualities can alienate those around us who don't like having someone questioning everything. Their worlds are comfortable precisely because they don't constantly question their surroundings.

As if things are not bad enough, it is almost impossible to tell a non-mathematician what we are doing. They don't have the patience or blackboard space to contain the twelve definitions we need to begin the discussion. And if we really try to explain ourselves, we just look even more abnormal to someone who cannot comprehend why anyone in this universe—or any parallel universe—could possibly derive excitement from this circumstance.

As the mathematician Janet Tremain puts it:

Intelligence is maladaptive.

3 An Explosive Subject

A quote of unknown origin goes:

A clever person solves a problem. A wise person avoids it.

3.1 Entering Nobel's Mind

Alfred Nobel died on December 10th, 1896 leaving the major part of his vast fortune to fund the Nobel Prize, which was designed to reward "science, literature and the quest for peace." It was to be given to those who "shall have conferred the greatest benefit on mankind." Mathematics was not honored with a prize. For a long time mathematicians speculated about why there was no Nobel Prize in mathematics. The reasons went from the ridiculous to the sublime. For an area that lives off truth, it was surprising how easily we were able to distort historical truth. Actually, we have no idea why Nobel did not have a prize in mathematics. I personally believe in the view given by Gårding and Hörmander: "The true answer to the question [why there is no Nobel Prize in Mathematics] is that, for natural reasons, the thought of a prize in mathematics never entered Nobel's mind."

3.2 A Field of Dreams

At the 1924 International Congress of Mathematicians a resolution was adopted to create medals to recognize outstanding mathematical achievement—later to be called the Fields Medal after J.C. Fields, the Secretary of the Congress. The Medal is awarded every four years on the occasion of the ICM to recognize "outstanding mathematical achievement for existing work and for the promise of future achievement." I guess, to leave adequate room for "future achievement," the Medal is awarded only to mathematicians below the age of

40. To here all was fine. We have an award for the best of the best young people to recognize their mathematical talents. But this all went awry when mathematicians started referring to this as the mathematical equivalent of the Nobel Prize. This left us looking foolish and ridiculous outside the field when we claimed to have the equivalent of the Nobel Prize, but it is not given for the most significant achievements in mathematics, but rather for the *youngest most significant achievements*. Unfortunately, this mistake has worked its way into Wikipedia and a number of other popular outlets.

3.3 Exploding With Success

The International Mathematical Union seems incapable of addressing the issue of a Nobel Prize in mathematics. Some people have suggested, for example, using the Abel Prize. The only saving grace here is that a new Nobel Prize in Economics was added in 1969 and so we can hope for our turn one day.

By the way, Alfred Nobel was a Swedish chemist and engineer. A series of disastrous accidents in his lab left a number of people dead (including his youngest brother Emil) while he tried to learn how to stabilize nitroglycerine. Eventually succeeding, he called his discovery *dynamite* and it was this invention that generated the massive fortune he used to fund the Nobel Prize. Nobel himself described 1860 as the time when he "made nitroglycerine *explode with success*."

4 The Goalkeeper

During World War II British Prime Minister Winston Churchill explained how to deal with severe adversity:

If you are going through hell—keep going.

When Jean Bourgain was just starting his career, Janet Tremain (a student at the time) asked him what his goals were. He said with his dry smile (something that the greatest code breakers of all time could not even begin to decipher): "to win a Fields Medal (the highest level award available at the time); to be at the Institute for Advanced Study (the most prestigious job in mathematics); and to make a lot of money (tricky, but not impossible in mathematics)."

4.1 Scoring Goals

When you enter mathematics, you will have to set your goals for your career. If you set your goals too high, you will spend your entire career frustrated and unhappy that you can't achieve what you want. We have enough stories in mathematics already about mathematicians ending their careers bitter and angry that they were not able to live up to their own unrealistic goals. For most of us, all we can hope for is to become "a first rate–second rate mathematician." This is already a lofty goal. There are just a tiny number of first rate mathematicians. Luckily, an army can't move forward if it consists only of generals. It takes a broad spectrum of mathematicians with all kinds of different talents to propel the subject forward. Also, the most critical need in mathematics is for truly creative ideas—and these can come from anyone.

4.2 Today is Yesterday's Tomorrow

Keep in mind also that for most of your career you will be "in progress" working on a project. If you only get enjoyment from major victories, you will have very few moments of happiness. You need to learn to enjoy the *process* of doing mathematics so that you can enjoy every day of your career. Yes, mathematics can be the most frustrating endeavor in the universe. It gives up its riches very grudgingly. And since we have declared this as our profession, we cannot be satisfied without rewards. But we can still enjoy the process of discovery even on a frustrating day when we make *negative progress*. That is a day when we realize that what we thought we did yesterday is false. But we can't move forward without that small insight. When asked once how his day was going, the cartoon figure Charlie Brown responded:

I keep hoping yesterday will be better.

4.3 Moving the Goalposts

Mathematics can be the perfect partner—fulfilling your every need. Or it can be your greatest nightmare—frustrating you at every turn. You are the *goalkeeper* of your career. If you tie your ego to your success, you will end up as a ring without identity.

By the way, Bourgain reached all three of his "goals" and much more (such as becoming a member of the National Academy of Sciences) at a very young age. So be sure not to set your goals too low. Or at least be prepared to move the goalposts as you go along.

5 Group Theory

Since your career in mathematics might be lengthy, please keep in mind the fundamental rule for long term group interaction:

Friends come and go, but enemies accumulate.

5.1 Simple Groups

When I went to my first mathematics meetings I noticed that the "stars" all hung out together, ate together etc. I assumed that we were not to disturb *the group* until we had built up enough *mathematical currency*. I later learned that this was not necessarily the case. Over years of being in a subject, other mathematicians may be our longest lasting friends. We have shared their marriages, raising of children (bunnies in my case) and the ups and downs of life. So we are glad to see old friends and catch up on what is going on with them. This is not designed as a slight on all the people we don't know so well. Getting into the group may not be a completely simple matter. But in my groups, at least, it is semi-simple. So give it a try.

5.2 The Inverse Elements

If things don't work out as you expected, don't let this discourage you. At my first math meeting I stopped *Mr. Big* in the hallway and said: "Hi. I am Pete Casazza. Would you have time to answer a question?" The response was: "Who are you?" Since I had a name tag and had just introduced myself, I quickly realized that this was *mathematics speak* for "Why are you important enough for me to talk to?" Since it was my first meeting, I concluded that overrating one's importance in the universe was a necessary condition for being a mathematician. Actually, such an experience is rare in mathematics and nothing near it ever happened to me again. But be prepared to meet a very broad spectrum of egos during your career.

6 Mathematics is Ageless

The comedian Jack Benny once said:

> *Age is a question of mind over matter. If you don't mind, it doesn't matter.*

6.1 The Age of Reason

Mathematicians have a paranoia about age. We are told from our earliest days that mathematics is a young person's game; that mathematicians do all their best work before the age of forty. This paranoia is made worse by the fact that the Fields Medal (see Section 3) has to be won before the age of forty. For a group that lives off logic, it is difficult to see how they so desperately hold onto this idea despite being faced with a very large number of counterexamples amongst us. Under this system, we should hand someone a Fields Medal for being the best and the brightest of the younger generation and say: "Here is your medal. By the way, your career is over."

6.2 Our Silver Anniversary

At one time, we would hold a special conference to honor a mathematician's 70th birthday. This made sense because that was often the forced retirement age. Today we have broadened our scope to honor the 60th, 65th, and 70th birthdays. But there is a segment of the community that is afraid to have this "honor" since it is tantamount to announcing the end of your career. If this age paranoia continues, then I would suggest we stop having such age-related honor meetings. Instead, why not have a special meeting for the 25th (30th etc.) anniversary of someone's entering mathematics? This would, of course, be your *silver* anniversary. But it would be better if we could just face this whole topic realistically in the first place.

7 Workman's Compensation

The famous actor Will Smith claimed that he had only average talent. That all his success stemmed from hard work.

> When the other guy is sleeping, I am working. When the other guy is vacationing, I am working. When the other guy is making love, I . . . uh . . . well, I am doing the same. But I am working really hard at it.

7.1 Input verses Output

As a student thinking about entering mathematics, don't be intimidated by its mystique. You do not have to be a genius to be a mathematician with a successful career. It helps to be of above average intelligence, but the most important tool at your disposal is *hard work*. Mathematics is not a *sprint* but rather a *marathon*. Hard work over a long period of time will pay off. It has been noted that today a large percentage of all gifted students severely underestimate their abilities. This comes from underrating the importance of effort. The system designed to help them is actually working against them. They are praised for being gifted which is something out of their control and therefore has limitations. When praised for their hard work instead, gifted students usually set higher standards, take more risks to succeed, and expect more of themselves. As Tom Lehrer once said:

> *Life is like a sewer. What you get out of it depends upon what you put into it.*

If you put your best efforts into mathematics you have a reasonably good chance for a successful career.

7.2 Some Inspiration

Fields Medalist Terence Tao put work in its place quite eloquently:

> The popular image of the lone (and possibly slightly mad) genius—who ignores the literature and other conventional wisdom and manages by some inexplicable inspiration (enhanced, perhaps, with a liberal dash of suffering) to come up with a breathtakingly original solution to a problem that confounded all the experts—is a charming and romantic image, but also a wildly inaccurate one, at least in the world of modern mathematics. We do have spectacular, deep and remarkable results and insights in this subject, of course, but they are the hard-won and cumulative achievement of years, decades, or even centuries of steady work and progress of many good and great mathematicians.
>
> Actually, I find the reality of mathematical research today—in which progress is obtained naturally and cumulatively as a consequence of hard work, directed by intuition, literature, and a bit of luck—to be far more satisfying than the romantic image that I had as a student of mathematics being advanced primarily by the mystic inspirations of some rare breed of "geniuses."

We have an expression for this in mathematics (which is based on a quote of Thomas Edison):

> *Success in mathematics is 1% inspiration and 99% perspiration.*

8 A Confidence Game

When Nobel Prize winning Physicist Albert Einstein's father asked the school principal what vocation his son should choose, the response was:

> *It doesn't matter, he'll never succeed at anything.*

8.1 The Advantages of Mathematics

One of the greatest challenges to your career will be to maintain your confidence—without being over-confident. Everything around you will be constantly testing your confidence. You are working on problems that appear unsolvable. You are constantly being evaluated for grants, jobs, promotion, tenure, raises etc. Deciding who will be the main speakers at meetings will be an evaluation process. Even when you achieve a great victory in your research, the first question that arises is: "Can I top this?" No matter how good you are, there is *always* someone better out there. There will be people around you who are faster, more knowledgeable, and more creative than you. This does not mean that you don't belong in mathematics. As T. Tao put it:

> This is the common mistake of mistaking *absolute advantage* for *comparative advantage*.... As long as you have education, interest, and a reasonable amount of talent, there will be some part of mathematics where you can make a solid and useful contribution.

8.2 In Praise of Mathematics

You will have to maintain your own confidence. Mathematicians are very stingy with praise. You will likely never hear someone say: "That was a great theorem. Thank you for bringing it to us." Or "Your book has greatly improved my mathematical life." Unfortunately, this is not part of the psyche of mathematicians—but it *should* be. *For some [natural?] reason, the thought of praise never enters mathematician's minds.*

8.3 True or False Questions

One of the most difficult problems in mathematics is learning how to balance *proper* respect for our *significant others* while maintaining a *healthy* respect for ourselves. If your whole measure of human beings is their mathematical achievements, you will be constantly undermining your own confidence. I often admonish my students for talking to themselves in the negative: "That was a dumb statement" or "That was really stupid on my part." I believe they became accustomed to doing that as a defense mechanism. If they say it first, it takes away the opportunity for others to say it. But every psychological study around shows that how we *talk to ourselves* is being heavily recorded in our subconscious and is forming our view of ourselves. Mathematical statements are not *smart or dumb*. They are only *true or false*.

You cannot afford to lose the *Confidence Game*: "A swindle in which the victim is defrauded after his or her confidence has been won."

9 You Can't Outrun a Bear

Einstein once said:

> *The definition of insanity is doing the same thing over and over again and expecting different results.*

9.1 Wonderful Advances

When I first joined the mathematics community I was excited to join a group dedicated to *advancing mathematics*. I had a rude awakening when it became clear that we were really working to *advance ourselves*. This is an unfortunate consequence of the reality around us. We must all compete for very scarce research grants, positions, promotion, tenure, awards, raises etc. But we need to be careful that this reality does not diminish our enjoyment of the subject. We need to be able to go to meetings and be excited and overjoyed at some wonderful advances—*done by someone else*. One always goes away feeling a little bit behind, but this is one of the main functions of meetings. They infuse us with added energy and drive to go home and do something serious.

9.2 A "Non-Profit" Organization

Mathematics is going through some difficult times at the moment. We have a shortage of jobs, low salaries, much of our funding is disappearing as the Defense Department shifts funds from mathematics and the National Science Foundation is funding only 10%–20% of the submitted proposals. Even NSF's figure is quite understated since more and more mathematicians are not even applying to NSF because the chances of being funded are so slim. At this time, the NSF budget is about $6 billion—approximately what the government will spend in 12 hours. Apparently, no one told them that serious researchers put in a 28 hour work day. An added problem is that even if NSF gets an "inflationary raise," the research they can support diminishes since they are supporting the most significant researchers in the sciences and engineering who are getting significantly above an inflationary raise. Apparently, the U.S. Government sees us as a *not for profit* group. They seem to have forgotten that this country has reached its current status by being the world leader in research. And while they consider increasing research funding to deal with the monumental problems facing the U.S. and the world, they are clueless that these problems would not even be here if they had adequately funded research in the first place. Worse, as they *target* research money to what *they* believe will give immediate relief to our problems, they will continue to underfund the futuristic research which represents the real long term future of the country—leaving us to wait for the next crisis we are not prepared for. This brings to mind an old proverb:

If two wrongs don't make a right—try a third.

9.3 Assisted Suicide

This is the right time for us all to come together for the good of the subject. We have three major math societies in the U.S.: The Mathematical Association of America (MAA) representing students and teachers of mathematics; the American Mathematical Society (AMS) representing *pure* mathematics; and the Society for Industrial and Applied Mathematics (SIAM) representing *applied* mathematics. Since I worked for 25 years in pure math and then switched into applied math and have been an active participant in the MAA, I have had an opportunity to witness all these groups in action. The pure math group looks down on the applied math group claiming they are not doing *serious* math. The applied math group looks down on the pure math group claiming they are developing deeper and more isolated theories that not only separate them further from applications but are isolating themselves even from other areas of pure math. Both of these groups have a certain level of disrespect for the "teaching wing." This is somewhat ridiculous since most of us are employed as teachers and doing research is desirable (or even required) but not viewed as our most important function by state legislatures.

At one time, all of mathematics grew out of applications. To establish its identity, it was natural that mathematics would separate itself from applications to build an independent future. But now it is time for us to come together. When I switched into applied math, the first thing I discovered was that some of the most important questions in pure math were not being addressed because they did not show up naturally there. They only showed up when one tried to apply the theory. Given the significant challenges facing mathematics, it is time for all the societies and mathematicians of all persuasions to come together for the good of the subject. But our natural inclination seems to be to compete instead of cooperate. Mathematics has become its own worst enemy—*assisting in our own suicide.*

9.4 Primitive Ideals

To understand all of the *apparent* contradictions above, I need to recall a story we used to tell in the boy scouts.

Two boy scouts are out hiking when they see a bear charging at them. One scout sits down, takes his running shoes out of his backpack and starts to put them on. The conversation goes:

> First Scout: *Why are you putting on running shoes? You can't outrun a bear.*
> Second Scout: *I don't have to outrun the bear. I only have to outrun you.*

10 A Black Cat That Isn't There

A quote sometimes credited to Einstein goes:

> *If I knew what I was doing, it wouldn't be called research.*

Mathematics differs from the other sciences in that we are attempting to *capture truth* while other sciences are trying to *approximate truth*. We also differ from the other sciences in that new discoveries don't falsify the old ones but instead extend what was known to capture a *broader truth*. What changes over time is our understanding of the mathematics—what it means and how it fits into the broader picture.

A story in the mathematics community is used to explain the difference between a mathematician and a physicist. Physicists work for ten years on a difficult problem and when they are done they say: "I am a genius for figuring this out." Mathematicians work for ten years on a problem and when they are done they say: "I am an idiot. The answer was obvious." Although this is exaggerated—as is this entire article—it does contain a shred of truth. Most mathematics, once completely uncovered seems somewhat obvious. But this should not be used to downgrade the enormous effort that went into it the first time.

As Charles Darwin put it:

A mathematician is a blind man in a dark room looking for a black cat which isn't there

11 My Most Read Paper

A famous review of a published math paper reads:

The results in this paper are false. The mistakes are not new.

11.1 Name Dropping

When you enter mathematics, you hope to produce that significant manuscript that will identify you forever as a major player in the field. We actually associate many mathematicians with their most significant contributions. Andrew Wiles will forever be attached to his solution to the 300 year old problem affectionately known as: "Fermat's Last Theorem." But this system also has its drawbacks. Max Zorn will forever be revered in mathematics for his *discovery* of *Zorn's Lemma*. Unfortunately, he made this discovery in his Ph.D. thesis and our association negates his whole illustrious career after that. But the fact is that few mathematicians will ever reach that level of significance and recognition. You will need to be satisfied with a name tag at meetings.

11.2 The Citation Index

You will have to develop enough confidence in yourself to be comfortable around people who just assume they are smarter than you. Otherwise, you will constantly be trying to broadcast your achievements to build up your own confidence. In my department, just in case you missed the significance of someone, they feel compelled to remind you with a complete lack of subtlety: "In my recent paper in the *Annals*, I showed...." This is *mathematics speak* for: "I am important since my last paper appeared in the highest level pure math journal." Not to be outdone, I cannot resist tooting my own horn and telling them about:

My Most Read Paper

One time I received an urgent message from a journal saying that I was holding up publication since I had not returned the galley proofs of my article. I replied that I had never received them. They resent them overnight mail and I managed to get them back quickly. Two months later, the original galley proofs arrived in a completely mangled package with multiple wrappings, tape and string. Inside was a disk that contained the TeX file of my article. On the various levels of covers it was clear that my disk had racked up a large number of frequent flyer miles. Its first trip was to Colombia, South America. Apparently, someone at the post office decided that Columbia, MO (my actual hometown) was in South America. The next stop for the manuscript was Venezuela followed later by Argentina. The last two addresses were in Washington, D.C. It isn't hard to figure out what happened. This strange disk with all its TEX symbols and large number of $dollar signs$ ended up in the country of Colombia where they quickly realized that this was something of real significance. Clearly these strange symbols and the very large number of dollar signs represented the entire operation for the largest drug cartel in the area. All they had to do was put enough agents on it to decipher it. Failing this, they enlisted the help of the secret service in Venezuela who certainly would decipher it. After passing through Argentina, there was only one group left who had a chance of breaking up this drug ring and that had to be in Washington, D.C. By my estimate, untold hundreds of dedicated law enforcement officials went through this manuscript with a fine tooth comb. This certainly has to be my most read manuscript—even if it is a little short of citations.

12 The Real Beauty of Mathematics

Author unknown:

The difference between genius and stupidity is that genius has its limits.

12.1 An Intelligence Test

Many mathematicians think of themselves as being more intelligent than the general public and even other scientists. Such a conclusion requires one to change the definition of *IQ* which is basically just *that which intelligence tests measure*. Intelligence tests do not even measure any non-trivial mathematics. They do measure elementary *logical reasoning* that most scientists use but only mathematicians have turned into a god. But there is no evidence supporting our innate belief in our superior intelligence. Certainly, one needs a certain amount of intelligence to work in mathematics. But we feel compelled to define *intelligence* as *being good at mathematics*. This is what allows faculty to sit in the lounge talking about *how stupid the students are.* In whose dictionary is "not being good at mathematics" the definition of stupid? That is, one does not have to be good at mathematics to be intelligent. The problem is that when one devotes one's whole life to a subject, one naturally begins to believe this is the only important thing to do. Grothendieck was one of the greatest minds of the last century. His manuscripts kept mathematicians busy for forty years just trying to fully understand this oracle. One day Grothendieck walked away from mathematics and became a farmer. This sacrilege was an unending topic of conversation forever at meetings since mathematicians just could not comprehend that someone so good at mathematics would choose not to do it. Clearly, if you are brilliant at math, you *must* do it.

12.2 Real Talent

It is difficult to work in mathematics without developing a certain amount of arrogance. Arrogance stems from seeing ourselves as better, smarter, and more intelligent than others—both inside and outside of mathematics. But this is not a winning long-term strategy. It only works as long as *others* are the idiots you think they are. But when someone significantly better than you comes along (*and they always do come along in mathematics*) you will spend the rest of your career trying to raise arrogance to the level of an art form so it will hide your clinical depression. There is a long list of famous mathematicians who ended their careers bitter and unfulfilled. It is easy to be arrogant. But it takes real confidence, intelligence, and talent not to be. Arrogance will not prevent you from being a significant mathematician, but it will remove 50% of the rewards—which involve your interactions with a spectacular group of dedicated mathematicians.

Perhaps the mathematician Janet Tremain summed it up best:

> *The real beauty of mathematics*
> *is that*
> *you don't need an especially high IQ to do it.*

13 An Eating Disorder

Professional baseball player Yogi Berra was asked by a waiter if he wanted his pizza cut into four pieces or eight pieces. He replied:

> *Four. I don't think I can eat eight.*

Going to dinner with mathematicians is a culinary experience from the *fourth dimension*. Only mathematicians could turn dinner into such a level of competition that it would qualify as an *Olympic event*. It is a four hour ordeal which *sometimes* even includes 30 minutes of actual eating.

13.1 We've Got to Lower Our Standards

Our first job is to pick the restaurant. This is where wisdom and experience pays off. Having spent enough meals denying that I had anything to do with the selection of this "god-awful place," I play *silence of the lambs*. As the experts work out the definition of *a good restaurant*, the rest of us are recalling the last time we interfered and thought we might have heard: "When I want your opinion—I will give it to you." After 30 minutes, we have finally arrived at a consensus—that we are hopelessly deadlocked and getting very hungry. We decide that our only choice is to walk down the street and see what is available. After walking far enough for the person behind me to work out a new *stopping time algorithm* (I wish I had put on my running shoes) and reading enough menus to fill the entire *conference proceedings*, we have come to agreement on a place. Unfortunately for us, by the time we all read the menu, the place closes. We realize that we have no choice but to *lower our standards* while a few restaurants are still open. We decide to take the next open place, whatever it is.

13.2 A Seat at the Table

Entering *Joe's: Eat At Your Own Risk* restaurant, we proceed to the ancient ritual of *correct seating order*—which is determined by one's *significance* to the mathematics community.

Basically, this amounts to organizing an elephant stampede into the order of importance of the elephants. This can make *musical chairs* look like child's play. There are more than 87 billion different ways 14 people can sit around a table and these guys are determined to try every one. Unfortunately, allowing 1 minute to try out each new position, it would take more than 160,000 years to exhaust all possible ways of seating 14 people around a table. It is best if you just sit at the bar and discuss algebra with the bartender until the dust settles. Anyway, you are at least guaranteed a seat this way.

13.3 Stop Global Whining

Next is the *wine tasting competition*. This has several different events such as: *Who knows the best wines? Who knows the most expensive wines? Who tasted the most wines on the list—and where? Who drinks only wines from their own country—and why?* Some people are already begging me for the glass of wine I brought from the bar. Finally we get through the last part of the competition: *Who can send back the most bottles of wine?* It is over except for assigning grades to the winners: It is too young, too warm, too cool, too spicy, doesn't go with the meal . . . I give my two cents: At least the cork tastes good and the label is pretty.

13.4 Extreme Sports

Dinner discussion for mathematicians raises competition to a higher plane. We are compelled to establish some sort of mathematical pecking order: the best, the worst, most creative, most potential . . . Then we need to relive the same dimensions with side conditions: biggest drinker, oldest, youngest, most arrogant, best driver, knows a lot about . . . (Did I just hear the name Grothendieck mentioned?). We feel compelled to identify all the extremes. After all, being average at anything is the greatest insult we can assign. Didn't we have this discussion last year? Who won?

13.5 A Taxing Situation

Finally, the waiter arrives and asks: "Do you want separate checks?" No thanks, *we can do the math*. Everyone starts talking amongst themselves computing what and how much or anything shared they had. We wind our way through bartering, trading, assigning weights for students, postdocs, untenured professors (why not emeritus?) grant supported or not, tax/tips etc. Running out of napkins to write on, someone is using the tablecloth as a calculator while others stare at the ceiling or their own feet trying to concentrate on the calculations. Each of us finally gets our bill computed to the exact cent. *Who says we can't balance a checkbook?* Unfortunately, everyone has only $20 bills. It is time to start making out the I.O.U.s. Does anyone have any paper?

As we get up from the table to leave, I make my first serious mistake of the evening (there goes my "no-hitter") and ask: "Does anyone know the best way to get home?" . . . God I wish I never said that. There's 30 minutes of my life I will never get back.

14 The Introduction—Finally

This article is intended as a survival guide for those students, teachers, and mathematicians who are having trouble interpreting the mathematical experience. If you read only this article, you will get a distorted image of the overall situation in mathematics. A comprehensive view of the subject would fill a textbook—preferably one on abnormal psychology. In lieu of this, the MAA has put out this book. Therefore, this article is purposely representative of nothing but my own personal experiences during 37 years of being a mathematician. To get a better view of mathematics and mathematicians, you will need to read the other carefully crafted articles in this book. I strongly recommend you visit Terence Tao's web page for more wisdom on these topics than any mathematician should ever be allowed to possess. Each issue raised in this article needed much more discussion. But my goal was only to raise these topics as items that the mathematics community needs to address. My long list of personal opinions on them is not particularly important. It was just important that I make them controversial enough to stimulate the much needed discussion.

If my article annoys you so much that you feel compelled to speak out—please do so since it means that I have done my job. This book is not designed to be the beginning of the end of the story but rather the end of the beginning. My satirical approach to this task is my response to the *sad fact* that no one has ever accused mathematicians of not taking themselves seriously enough.

15 The Last 10 Minutes

Being a mathematician is the greatest job in the world. Every day is even more exciting than the last. Each day starts with mental gymnastics. And today has the potential of being the day that you finally crack that tough nut. It is a challenging, very stimulating, treasure hunt. It is creative brainstorming at its best. Most of the non-mathematicians my age I know are already burned out and are trying to hang on until retirement. But even after 36 years of doing mathematics, I go to bed at 2 PM in the afternoon and get up at 10 PM so I can get to the thing I love as early as possible. This of course occurred because of my obsessive-compulsive nature. I started getting up at 5 AM so I could get two hours of research done before going to the university. Then I figured out that getting up at 4 AM gave me an extra hour. After a short time I was getting up at 10 PM. Pretty soon I won't have to go to bed since I will be getting up at that time. The point is, few people will have the opportunity to have a career that is so constantly exciting, rewarding and challenging that they cannot imagine retiring from it. When you get into it you will realize that a day without mathematics is a day without sunshine. Or as a famous expression goes:

Mathematics is not a matter of life and death.
It is much more important than that.

So if you are not yet one of us, come join in. Lighten up, chill out, relax... it's just mathematics and you are just a person. Enjoy the treasure hunt.

Someone asked me once if I planned on doing mathematics my whole life. I gave the obvious answer:

Of course not. I plan on saving the last 10 minutes to reminisce.

Acknowledgment. I fretted for almost a year over how I was going to write (or even approach) this article. Then I attended the GPOTS conference at the University of Cincinnati and spent the week carefully discussing this article with my friends and colleagues. It soon was completely clear what I needed to write. Returning home, I wrote the first draft of this article in two days. I am indebted to everyone who attended the meeting for all their insights and for just being their wonderful selves. A special thanks goes to Don Hadwin for supplying the creative title for Section 2. Many others sent much needed improvements: Karlheinz Gröchenig, Chris Heil, Norbert Kaiblinger, Jelena Kovačević, Sergei Novikov. A large portion of the material for this article came from Janet Tremain's eidetic memory. I sat for hours while she played video tapes stored in her mind recounting minute details of events in our past interactions with mathematicians. One example: she replayed a dinner we had twenty years ago with a group of mathematicians, describing where each person sat at the table, what each person ate, who sent back their steak to be cooked more, what wine/beer each person drank, and each conversation going on at the table. My response: "Janet, you are scaring me!"

5

We Are Different

Underwood Dudley

By "we" I mean those of us who earn (or, if retired, earned) our keep through mathematics. Mostly I mean those who prove theorems, though much of what I say applies to those who teach calculus, tell insurance companies what rates to charge, construct models to make hedge fund managers rich—things like that. By "different" I mean . . . well, *not the same* as the non-mathematical run of humanity. "Of course," you may think, "we know more mathematics." But there is more to it than that. I will point out some of the differences and give what I think are reasons for at least some of them. I will save them for later, especially because I will repeat them several times, and also because they are based on intuition and speculation. Until social scientists get around to conducting a parallel study of 2,000 mathematicians and 2,000 non-mathematicians over twenty or thirty years, speculation is all we have. However, speculation can be useful. The Riemann Hypothesis is speculation.

We can get at the differences seeing what the general public thinks of us. In the movies, for one place, we do not come out looking too well. Insanity is one of our main characteristics. In "A Beautiful Mind," a 2001 film about John Nash, the mathematician who won a Nobel Prize for economics, it could not be helped because its subject was for many years insane by any definition of the term. The hero of "Good Will Hunting" (1997) is not clinically crazy, but someone with immense mathematical talent who chooses to work as a janitor and isolate himself from life; he qualifies as being on the far reaches of eccentricity. He also has an amazing memory and is able to do mental arithmetic with great facility, common characteristics of mathematicians in popular culture. This is annoying if you are someone like me, whose checkbook never balances at the end of the month. It is also annoying that, in the film, the difficult unsolved problem that Will Hunting solves can be disposed of in a few lines. Proofs are seldom that short. Well, that's Hollywood.

"Proof," the good 2005 movie based on the even better 2000 play by David Auburn has as its center a father and daughter, both mathematicians. The father was completely gaga and doubts were raised about the daughter's sanity. She was, at the least, unusual. Why, if she had indeed solved an important problem, did she not submit her result to, say, the *Transactions of the American Mathematical Society*? Of course, if she had there would have been no movie. The protagonist of "Π" (1998), a mathematician, is crazy from the start and gets crazier. He can, as usual, do large calculations in his head. The movie is filled with nonsensical numerology and is, in my view, as insane as its hero who, at its end, drills a hole in his head, thereby finding peace at last. "La Habitación de Fermat," a 2007 Spanish movie, has characters called Galois, Hilbert, Pascal, and Fermat. Because it was a thriller (the mathematicians are trapped in a room whose walls were closing in) it was not to be expected that the characters would be behaving like ordinary people, which they didn't. One reviewer wrote, "And let's face it: math is scary all by itself, even without the mysterious parties, angry strangers, and a freaky shrinking room."

Yes, math is scary and so, by extension, are mathematicians. We are, if not completely fright-inspiring, *other*. The movies tell us so.

On television, the series "Numb3rs" had considerable success, lasting for six seasons from 2005 to 2010. The heroes were an FBI agent and his mathematician brother, who solved crimes using mathematics. The mathematician character did not look funny (as, we will see later, many people think that mathematicians do) and the mathematics was not preposterous. However, it was hard to distinguish the mathematician from the magician, waving a mathematical magic wand—hey presto!—to pull the rabbit, or the solution, out of the hat. The public view of the program was, I think, given by a one-line synopsis "Working for the FBI, a mathematician uses equations to help solve various crimes."

The best-known mathematician in fiction is probably Professor Moriarty, Sherlock Holmes's opponent in the stories by Arthur Conan Doyle. Moriarty had the "high domed forehead" that mathematicians in popular culture tend to have, necessary to contain his giant intellect. Doyle tells us that "At the age of twenty-one he wrote *A Treatise on the Binomial Theorem* which has had a European vogue." (Would that our papers could have vogues, continent-wide or not! Even in 1893, when the Professor made his first appearance, I do not think that mathematical work circulated very far outside the profession.) Moriarty also wrote *The Dynamics of an Asteroid*, with Doyle perhaps recalling Gauss's publications on the orbit of Ceres, though Gauss did not give any of them that title.

Moriarty left academic life under a cloud, moved to London, and turned to crime, becoming "the organiser of half that is evil and nearly all that is undetected in this great city." That is unusual. Mathematicians, being thought of as being unworldly, are not given credit for organizational powers. In academic life, when mathematicians turn to administration, they tend to go no further than upper-middle management, seldom rising to the top. Nor are they often found as CEOs of corporations, criminal or otherwise. Doyle made Moriarty a mathematician, I think, because Moriarty had to be very, very smart, and that is what mathematicians are.

In *The Bishop Murder Case* (1928) by S. S. Van Dine, a pseudonym of Willard Huntington Wright (1887–1939), many of those who are murdered are mathematicians, as is the murderer. Wright, who was an art critic before he was a detective story writer, somewhere picked up some mathematics. The book refers to the Riemann-Christoffel tensor and in chapter 21 his insufferable detective, Philo Vance, says that "Neither Newton nor Leibniz nor Bernoulli ever dreamed of a continuous function without a tangent." I conclude that Wright was not killing off mathematicians as revenge for unfortunate experiences in algebra class. Nevertheless, he made his murderer crazy, driven to murder by mathematics. Philo Vance is speaking:

> In order to understand these crimes, we must consider the stock-in-trade of the mathematician, for all his speculations and computations tend to emphasize the relative insignificance of this planet and the unimportance of human life.
>
> Markham, there's no escaping the fact: these fantastic and seemingly incredible murders were planned by a mathematician as forced outlets to a life of tense abstract speculation and emotional repression.

Mathematics, Wright and many other people think, pulls mathematicians away from the non-mathematical world. I believe this is so in many cases. Mathematics is much more interesting than what appears in the daily newspapers.

Books with mathematicians as characters continue to appear. *La Solitudine dei Numen Primi* (2008) by Paolo Giordano has sold more than a million copies in Italy and has recently been translated into English by Shawn Whiteside as *The Solitude of Prime Numbers*. The main characters, the male a mathematician, "Stand apart, outside ... trapped in closed circuits of self-involvement" as the New York *Times* reviewer, Liesl Schillinger, said (April 11, 2010). The female character knows that the mathematician is "'strange,' but so are most mathematicians, she rationalizes: 'the subject they studied seemed only to attract sinister characters.'" In *Uncle Petros and Goldbach's Conjecture* by Apostolos Doxiadis (2000) the uncle is "obsessed." And so it goes.

Mathematicians who are murderers can be found outside of the pages of murder mysteries. André Bloch (1893–1948) killed one of his brothers, an uncle, and an aunt in 1917 and was confined to an asylum for the rest of his life where he produced respectable mathematics, including a theorem that bears his name. In 1996 Walter Petryshyn (b. 1929), a professor at Rutgers, killed his wife with thirty-odd blows of a hammer. He had been exhibiting odd behavior, being upset about an error that had appeared in his most recent book. He had the delusion that it would make him a mathematical laughingstock, and the delusion that his wife was plotting against him. A year later he was found not guilty by reason of insanity. Theodore Kaczynski (b. 1942), the Unabomber who was responsible for three deaths, was captured in the same year. John Allen Paulos published an OpEd piece in the New York *Times* (4/7/1996) that contained similarities to Philo Vance's view of mathematics. A characteristic of the subject is its abstractness, he said, and though the ability to think abstractly is a precious one,

> Nevertheless, abstract thinking has been associated with various pathologies, and it is easy to see how one trained in such reasoning and in thrall to an ideal could come to justify murderous acts as a nebulous "good."
>
> Mathematics is also beautiful, but its aesthetic—minimalist, austere—can blind one to the messiness and contingencies of the real world.

My time as a graduate student at the University of Michigan overlapped Kaczynski's for two years, but as far as I was concerned he was an abstraction—I never encountered him.

On the whole, I don't think that mathematics predisposes to murder. The murder rate in the U.S. has declined to about 6 per 100,000 in recent years, so if there are 100,000 mathematicians in the country and they are unusually given to murdering they should be committing more than six murders a year. I do not think this is the case. In general, mathematicians are weird, but harmless.

We can get an indication of what people think of mathematicians by the jokes about them. Some are in-jokes, such as the one in which a mathematician is asked for an anagram of "Banach-Tarski" and replies "Banach-Tarski Banach-Tarski." Ones for the general public are more along the lines of the following. Don't stop me if you've heard this one, because you probably have. It illustrates the popular image of mathematicians as fussy precisionists, given to unneeded pedantry. An engineer, a physicist, and a mathematician were driving past a field that was populated by black sheep. The engineer said, "I see that all the sheep around here are black." The physicist said, "You can't say that. All you can say is that all the sheep in that field are black." The mathematician said, "No. The sheep in that field are black on at least one side."

The genre is large. Here is another example. The E, P, and M are confronted with a house on fire: what to do? The engineer said, "Call the fire department." The physicist said, after a quick calculation, "Apply 127,000 gallons of water." The mathematician said, "A solution exists." The jokes are funny because the mathematician is logical but disconnected from reality which, in the joke-teller's mind, is where mathematicians are. I will spare you jokes about mathematicians reducing things to previous cases and changing light bulbs.

People's attitudes towards mathematics and mathematicians are often formed in childhood and never changed. Susan H. Picker and John S. Berry reported in *Educational Studies in Mathematics* (vol. 43 #1 (2000)) on an experiment they had done: having students aged 12 to 13 draw pictures of mathematicians and say what they thought about them. They reproduced some of the pictures. The first is of a quite unattractive man on which the artist had included annotations to some of its features: "dirty unwashed hair, wrinkles from thinking too hard, an old stain—he's too lazy to wash his shirt, bad body posture, fat from doing nothing but math, pants are too small, hole in wrinkled pants—he's too lazy to buy new pants." Ah, to be seen as others see us! The artist may later see more mathematicians and teachers of mathematics and may alter his ideas somewhat, but I think for the majority what they report at age 13 is what they will report at 23, 33, or any later age. The authors had students in the US, the UK, Finland, Sweden, and Romania draw pictures. They showed that mathematicians are seen in the same way in all five countries. I was surprised that only a fifth of the pictures were of teachers, the only people familiar with mathematics that most participants in the study would have encountered; four-fifths were of what the students thought mathematicians looked like.

Most of the pictures of teachers were of nasty people using coercion, some armed with guns (though only in Sweden and Finland, a curiosity that deserves to be investigated), and one who had a devil's tail. They were intimidating powerless students ("You should all know this!"), who were often drawn considerably smaller than the teacher.

A common theme of the non-teacher pictures was that mathematicians are foolish. They have funny hair, their socks don't match, they "have no friends and wear high water pants and thick black glasses." A pupil in Romania wrote of her mathematician that "His clothes are out of fashion because for a long time he only stayed in his office." Another student gave a list of characteristics:

> have no friends (except other mathematicians)
> are not married or seeing anyone
> are usually fat from doing nothing but math
> are very unstylish
> have no social life whatsoever
> are 30 years old
> have a very short temper

The idea that mathematicians are alone and separate from the rest of humanity somehow becomes implanted at an early age. In general, though, we are *not* fat. Some of the other students' perceptions may also be slightly in error.

It is probably time for me to say why I think mathematicians exhibit the characteristics that they do. As Tom Lehrer said in his introduction to his song "Lobachevski," "Some of you may have had occasion to run into mathematicians and to wonder therefore how they got that way." There are two reasons. The first is that they *know* mathematics. The second is that they know *mathematics*.

Mathematicians are, as I said, different. For one thing, they know more mathematics than ordinary people. For another, they do not look like them. Inspect the line of people waiting to get into an airplane going to the city where the annual mathematics meetings are to take place the next day. You can pick out the mathematicians. They have the mathematical air, that combination of scruffiness and intelligence, of absent-mindedness and self-possession, not often seen among the sales representatives, corporate planners, babies, and others who travel in the cheap seats of airplanes.

Where does the air come from? I think that I know. It comes from mastering a subject and knowing that the subject has been mastered. None of us have mastered the whole of mathematics, of course, but knowing that you have an up-to-date grasp of, say, uniform distribution modulo one, and of all the topics in the courses that you teach your students, entitles you to the inner confidence, seen also in athletes and musicians, that lets you be scruffy if you so choose. The opinions of other people, especially those of other people on airplanes, are not all that important.

We *know* mathematics, and we know that we know it. This benefit is not available to practitioners of other disciplines. Physicists might like to prove things, but they can never be sure that they have seen nature whole. Something may turn up that will force them to alter what they have established, as happened when Einstein revised and extended Newton. So, arrogant as some physicists can be, they are not entitled to the serene sureness of mathematicians, who, having proved something, need never fear that it will be superseded. Generalized, maybe; replaced, never.

Philosophers can never prove a theorem because as soon as they try, some other philosopher will explain how it is the result of muddled thinking. Philosophers can never reach a conclusion. They always have to be looking over their shoulders, seeing who is sneaking up on them, and thus can never have the confidence of mathematicians. Even worse off are the members of the English Department, where it can be debated if there even is a discipline. Professors of literature can never think that they have mastered anything. In a feeble imitation of mathematics, the poor things have been reduced to constructing "theory" with which to occupy themselves. Only mathematicians have behind them the solid rock of proof and a body of knowledge that they *know*.

The second reason is that we know *mathematics*. Consider what Howard Eves had to say in his *In Mathematical Circles* (1969, reprinted by the Mathematical Association of America, 2002):

> Having associated from early years with two classes of scholars—botanists (or, more widely, nature lovers) and mathematicians—I came to notice, and through the years have confirmed, a striking difference between the two classes. The botanists are usually the most pleasant sort of people to be with; they radiate gentle modesty, are open-minded, enjoy each other's company, are kind in their professional comments about one another, and are found interesting by their non-botanical friends. The mathematicians, on the other hand, are too often unpleasant to be with; they frequently exude self-confidence, are professionally opinionated, tend to bicker and quarrel among themselves and say unkind things about one another, take an almost gleeful pleasure in unearthing an error in another's work, and are quite often boring to their nonmathematical acquaintances.
>
> The unpleasant features of the mathematical group are noticeable even among some of the more gifted high school students of the subject, become sharper among the college graduate students of mathematics, and often attain an undignified aspect among the college instructors and professors of mathematics.

Now why is that? Some of it may be due to genetics. There was a study by six authors that appeared in *Autism* in 1998 of students at the University of Cambridge, 641 in mathematics, physics, and engineering and 652 in literature, either English or French. They were asked how many of their relatives suffered from various maladies, as schizophrenia or Down's syndrome. There were two statistically significant results. Science students were six times as likely as literature students to have a relative with autism, and the literature students were twice as likely as the science students to have a relative with bipolar disorder. So, mathematicians seem to be more likely to carry the gene, or genes, of autism. Victims of autism share, in an exaggerated degree, some of the characteristics of mathematicians, as lack of empathy and preoccupation with certain topics.

As Stanislaw Ulam said in his *Adventures of a Mathematician* (1976),

> In many cases, mathematics is an escape from reality. The mathematician finds his own monastic niche and happiness in pursuits that are disconnected from external affairs. Some practice it as if using a drug. Chess sometimes plays a similar role. In their unhappiness over the events of this world, some immerse themselves in a kind of self-sufficiency in mathematics. (Some have engaged in it for this reason alone.)

Genetics aside, I think that the subject of mathematics must bear some of the blame. When a botanist makes a discovery, his or her reaction could very well be, "Look at that! Isn't nature wonderful?" When a mathematician proves a theorem, his or her reaction, conscious or unconscious, could very well be, "Look at that! Am I not wonderful?" Botanists can be made humble by the natural world, mathematicians cannot. The botanist is an observer, the mathematician a creator—how like a god! There is nothing in mathematics to induce humility; all the force is in the other direction, towards pride and arrogance. Also, many mathematicians earn money by teaching and there is none of the humanists' learning from their students in mathematics: teachers of mathematics know the subject, students do not; the teachers have the right answers, all of them, and students will never win a mathematical argument with them (not that any student would try). It is not hard to make the transition from being always right in the classroom to being always right everywhere.

In *A Mathematician's Miscellany* (1953) J. E. Littlewood said "Mathematics is a dangerous profession; an appreciable proportion of us go mad." I don't think he had that quite right. Some of us may seem mad, but we are not. Paul Erdős was fearfully eccentric, so much so that a superficial observer might put him down as a nut, but that would be an error. I'm fairly sure that he knew that he was eccentric but he did not let it bother him, and if it bothered others, well, so much the worse for them. Mathematics is what was important. Mathematics trumps eccentricity.

Perfect Rigor (2009) by Masha Gessen is a book about Grigori Perelman, who verified Poincaré's conjecture and declined the million dollar prize that he could have collected. When asked if he was crazy Gessen said "I think this is true of Perelman." He had refused to talk to her, but refusing to talk to journalistic pests may be more of a sign of good sense than of nuttiness. Not taking the money might seem to some to be the height of insanity, but Perelman had principles, perfectly good ones, that led him to that decision. Withdrawing from the world allows him to spend more of his time on mathematics. Because that is, I think, the best use of his time, he is proceeding in a reasonable manner.

Alexander Grothendieck also refused a prize and has given up mathematics. This does not mean that his mind is weak. I think, rather, that it demonstrates the strong-mindedness of mathematicians. They know what is important and they go after it, no matter what the rest of the world thinks.

Let us consider anecdotes about mathematicians. Most of them are about their smartness, their absent-mindedness, or their unpleasantness.

Everyone knows that mathematicians are smart. The United States Employment Service once tested people in various occupations for intelligence, manual dexterity, motor coordination, and so on. The highest score for intelligence, as measured by IQ score, was 143 for mathematicians, above dentist (131), accountant (118), waitress (80), and all the others. (The standard deviation of the numbers seems high, but they are certainly qualitatively correct.) That mathematicians are absent-minded is a corollary of the power of concentration that the discipline demands. That they are unpleasant is more difficult to explain though not more difficult to document.

Anecdotes about smartness go all the way back to Thales, the first mathematician. Remember the one about his cornering the olive-oil market? About telling the woman where to find her lost wash? About the mule? Even if you remember the last one, it is so good that it deserves retelling. Thales, the story goes, had a salt mine and mules were used to haul the salt to town for sale. The route crossed a stream; one day a mule slipped and fell in, the water dissolved most of the salt, and the load was much lighter. On the next trip the clever mule—oops!—fell in again. How to cure the mule of this bad habit? Yelling, beatings, moral suasion? Mules, anecdotes tell us, are stubborn and those methods would not be likely to succeed. Thales, being a mathematician, filled the mule's bags on the next trip not with salt but with sponges.

Search for anecdotes testifying to the warmth, generosity, or humanity of mathematicians and of course you will find a few. However, for every one you find you will find two, or twenty, testifying to the smartness, absent-mindedness, or unpleasantness. What are the anecdotes about Newton?

> Once, having dismounted from his horse to lead him up a hill, the horse slipped his head out of the bridle; but Newton, oblivious, never discovered it till, on reaching a tollgate at the top of the hill, he turned to remount and perceived that the bridle he had in his hand had no horse attached to it.

Another is

> On getting out of bed in the morning, he has been discovered to sit on his bedside for hours without dressing himself, utterly absorbed in thought.

These, and many others along the same lines, can be found in R. E. Moritz's *Memoribilia Mathematica*.

That book was published in 1914 (and was reprinted by the Mathematical Association of America in 1993), but similar anecdotes have been accumulating since then. Here is one about Norbert Wiener, as told by George Pólya in "Some mathematicians I have known," which appeared in the September, 1969, issue of *the American Mathematical Monthly*.

> It is about a student who had a great admiration for Wiener, but never had an opportunity to talk to him.

> The student walked into a post office one morning. There was Wiener, and in front of Wiener a sheet of paper on the desk at which he looked with tremendous concentration. The student was deeply impressed by the prodigious mental effort mirrored in Wiener's face. He had just one doubt. Should he speak to Wiener or not? Then suddenly there was no doubt, because Wiener, running away from the paper, ran directly into the student who then had to say, "Good morning, Professor Wiener." Wiener stopped, stared, slapped his forehead and said: "Wiener—that's the word."

As Pólya said, the story "is hardly true" but that is the case with a huge number of anecdotes. They are made up to illustrate truths that can be illustrated in no better way. The anecdote, true or not, is better than saying "Wiener was absent-minded."

Wiener also forgot where he lived, Newton forgot to eat, and Archimedes failed to notice a battle going on around him, but I will not add to the list of absent-minded anecdotes.

There is nothing bad in being smart and absent-mindedness can be endearing, but there is little redeeming at being unpleasant. Here is David Hilbert in action at a meeting of the mathematics seminar in Göttingen as reported by Constance Reid in her biography *Hilbert* (1970). The meeting room had two doors: one to the hall and another that led to the mathematics reading room. A visitor was lecturing on a topic in differential equations when

> ... [Hilbert] interrupted the speaker with, "My dear colleague, I am very much afraid that you do not know what a differential equation is." Stunned and humiliated, the

man turned instantly and left the meeting, going into the next room, which was the reading room. "You really shouldn't have done that," everyone scolded Hilbert. "But he doesn't know what a differential equation is," Hilbert insisted. "Now, you see, he has gone into the reading room to look it up!"

That this discreditable anecdote should appear in a sympathetic biography is testimony to Hilbert's unpleasantness, and to the lack of better anecdotes. Hilbert, you see, did know what a differential equation was, and, being Hilbert, did not see the need of wasting his time with a mathematical inferior.

Not to pick on Hilbert, but here is another anecdote from Pólya's reminiscences. It was the custom in Hilbert's Göttingen for new faculty members to make the rounds of his colleague's houses to introduce themselves. One came to Hilbert's house, put his top hat on the floor, and started talking. However, he talked too long, so

[Hilbert] stood up, took the top hat from the floor, put it on his head, touched the arm of his wife and said: "I think, my dear, we have delayed the Herr Kollege long enough."—and walked out of his own house.

That anecdote may be true. It sounds as if it would be hard to make up.

What do we know of Euclid as a person? Precisely nothing except for two anecdotes. One is about the member of the aristocracy who wanted to learn geometry, but wished to do it without going to all the work of plowing through the *Elements*. "There is no royal road to geometry," Euclid is supposed to have said. The other is about the student who asked, as students will, what it was all good for. Euclid gave instructions that the student be given a dollar (or the ancient Greek equivalent thereof) so that he could say that he had gotten something out of mathematics. The mathematician, ever the master of the put-down. Did you notice that he didn't even speak to the student? Even if false, the anecdotes give the impression that Euclid was cold, austere, a bit haughty, and by no means cuddly.

His characteristics live on in some mathematicians today. A 20-page "In Memoriam" article that appeared in the *Bulletin of the American Mathematical Society* some time ago (I will not name its subject because he was not a public figure in the sense that Euclid and Hilbert are) contained, as such articles must, human touches about its subject. "Some of his former students recollect his sense of humor" the author said, and gave two examples. One was his subject's arranging to appear twice in a composite photograph by running from one end of the group to the other between shots—funny enough. But the other is

A former student who felt himself a novice in teaching sought advice from [him] to which the latter gave the terse reply: "Always start writing in the upper left-hand corner of the blackboard."

Are you laughing? Remember, this is the *best* that the author could find to illustrate the deceased's warmth and sense of humor. The piece contained two other human touches. One was his habit of opening classroom windows whatever the weather was like (*he* was moving around, *he* wouldn't get cold) and the other was his throwing into the wastebasket at the end of each class the piece of chalk he had been using. "Besides," the author of the piece wrote, "during the run of a lecture he was occasionally known to have used an inattentive student as a target for the chalk." I am certain that the author was not indulging in irony, painting the portrait of a thoroughly disagreeable man; he was doing the best that he could. Or, being a mathematician, doing the best that he thought it necessary to do. Mathematicians are devoted to the truth and see no need to veil it with the pleasant semi-truth. It is one of the reasons why they do not go far in politics. Most mathematical obituaries concentrate on the deceased's mathematical work and go into personalities hardly at all, perhaps with good reason.

Steven G. Krantz has published two volumes of mathematical stories and anecdotes, *Mathematical Apocrypha* and *Mathematical Apocrypha Redux* (Mathematical Association of America, 2002 and 2005), having previously published an article on anecdotes in the *Mathematical Intelligencer* (vol. 12, #4, 1990). His purpose was not to discredit the anecdotees—Bergman, Besicovich, Gödel, Lefchetz, and Wiener. He said that "The enormous scholarly reputation of these men sometimes cause their humanity to be forgotten" and "In telling stories about them we bring them back to life and celebrate their careers."

What do we find about Bergman (1898–1977)?

> Bergman had always felt that the value of his ideas was not sufficiently appreciated.... [At a conference] I sat next to him at most of the principal lectures. In each of these he listened carefully for the phrase "and in 1922 Stefan Bergman invented the kernel function." Bergman would then dutifully record this fact in his notes—and nothing more. I must have seen him do this twenty times during the three-week conference.

That is one anecdote. Another is

> On another occasion a young mathematician gave Bergman a manuscript he had just written. Bergman read it and said "I like your result. Let's make it a joint paper, and I'll write the next one."

Then there is the story about how Bergman was going to leave his own wedding reception to discuss mathematics, until he was threatened with losing his job it he did. There also is

> Once he phoned a student, at the student's home number, at 2 a. m. and said "Are you in the library? I want you to look something up for me."

> Another time, one of his students got married and set off for his honeymoon, in California, by bus. Bergman was going to California for a conference and decided to take the same bus.

> The student protested that the trip was to be part of his honeymoon, and that he could not talk mathematics on the bus. Bergman promised to behave. When the bus took off, Bergman was at the back of the bus and, just to be safe, Bergman's student took a window seat near the front with his wife in the adjacent aisle seat. But after about ten minutes Bergman got a great idea, wandered up the aisle, leaned across the scowling bride, and began to discuss mathematics. It wasn't long before the wife was in the back of the bus and Bergman next to his student—and so it remained for the remainder of the bus trip!

I have included so many anecdotes about Bergman to avoid the accusation that I was selectively picking stories to put him in a bad light. Bergman must have been the person they describe: consumed by mathematics, self-centered, and taking no notice of the wishes of others

> One partial reason for the scholars' vices that mathematicians all too often display—pride, ungenerosity, spleen, bile, self-righteousness—is that mathematicians are unappreciated. It is not that they think they are unappreciated: they genuinely *are*.

As Alfred Adler put it in an article for non-mathematicians in the February 19, 1972 issue of the *New Yorker*

> Of course money and power are only superficial rewards, and it is possible to live without them; less tangible rewards, those of recognition and understanding, are what make effort and accomplishment rich and fulfilling when things are going well, and effort and failure bearable when things are going badly. But mathematicians cannot expect these either. For example, it would be astonishing if the reader could identify more than two of the following names: Gauss, Cauchy, Euler, Hilbert, Riemann. It would be equally astonishing if he should be unfamiliar with the names of Mann, Stravinsky, de Kooning, Pasteur, John Dewey. The point is not that the first five are the mathematical equivalents of the second five. They are not. They are the mathematical equivalents of Tolstoy, Beethoven, Rembrandt, Darwin, Freud.

There is no Nobel Prize in mathematics. The world is ignorant and uncaring. Is it not galling to know that no matter what you do, even if you settle Goldbach's conjecture and the Riemann hypothesis in the same year there will never be for you the praise and recognition that others get for accomplishments that are orders of magnitude less worthy? Does it not grate to know that all there will be is the grudging respect of colleagues—grudging because after they read your proof of the theorem they can see how obvious it was: they could have done it had they only thought of it—and perhaps a line or two in future histories of mathematics. As Adler wrote,

> In the company of friends, writers can discuss their books, economists the state of the economy, lawyers their latest cases, and businessmen their latest acquisitions, but mathematicians cannot discuss their mathematics at all. And the more profound their work, the less understandable it is.

It is a bleak prospect, which can build frustration, and frustration bottled up can change to sourness, envy, and jealousy. There seems to be no remedy.

Being unappreciated cannot by itself explain the characteristics of mathematicians. Botanists are also unappreciated (I could not name one American botanist) but Howard Eves assures us that they are pleasant people. Perhaps mathematicians' tendency to peak early plays a part. Gauss said that his accomplishments were only the working out of ideas he had had before he was 20. One of the bigger names in American mathematics said to me, when he was around 60, "I can only prove the trivial stuff now." As Adler noted,

> The limits of life grow more evident; it becomes clear that great work can be done rarely, if at all. Moreover, there are family responsibilities and professional sinecures. Hard work can certainly continue. But creativity requires more than steady, hard, regular, capable work. It requires total commitment over years, with the likelihood of failure at the end, and so the likelihood of a total waste of those years. It requires

work of truly immense concentration. Such consuming commitment can rarely be continued into middle and old age, and mathematicians after a time do minor work.

In botany and other disciplines, life is easier. Work will bring reward: enough time spent in the library, the laboratory, or the greenhouse, enough effort gathering data, enough work analyzing and summarizing, will give results. Practitioners who have only a modest amount of talent can do things that are useful. It is not that way in mathematics. Trivial problems are easy to recognize as trivial. It is possible to attack a non-trivial problem that is too hard and that you will never be able to solve because you lack the necessary talent, but you do not know that before you start. You do know it when you at last give up with nothing to show for it. It is a melancholy experience. Mathematics is stern, and unbending.

Nevertheless, here we are. We are born, many of us, I think, predisposed to like and be good at mathematics. We succumb to its lure and become mathematicians. The subject takes us over. It is important, more important than those ephemera that flutter up every day only to wither and be replaced by the next day's ephemera. Who cares about them? We take on, we are forced to take on, the characteristics that mathematics brings with it. We are different. We can't help it.

6

The Naked Lecturer

T. W. Körner

Now that I have your attention, I would like to to write about mathematics lecturing. If readers are annoyed by my confusing the pronouns "you," "one," "she" and "he," they should remember that I think of myself as addressing an audience of mathematicians and non-mathematicians. The non-mathematicians are rather shadowy creatures, but the mathematicians have clear characters and include both men and women.

1 Lecture Courses

For most mathematicians lecturing is part of their job. A few mathematicians grudge every moment of teaching as a moment taken from their research. Of course, teaching may occasionally aid research. Conway lectured on the construction of the real numbers starting with naive set theory, giving a different version of the standard constructions each year. I suspect that he would not have produced the theory of surreal numbers if he had not given those lectures. When I lectured on elementary functional analysis, I would always close the section on Baire's category theorem by saying that, almost certainly, many variations on the ideas remained to be exploited. Later I came across one such variation. Even so, it is reasonable to suppose that, in general, the more effort devoted to teaching, the less effort is devoted to research.

In spite of this, most mathematicians do not regret the time they spend teaching. In part, this reflects a feeling that, having been well taught ourselves, it is our duty to give back something of what we owe. In part, it reflects the fact that research is, on the whole, a lonely occupation and teaching is a social one. In Eastern Europe under communism, politically suspect mathematicians might be moved from teaching universities to research-only institutes with the move intended as a punishment and felt as such. Mainly, I think, mathematicians like to lecture for the reasons outlined in *Surely You're Joking, Mr. Feynman!*. If you only do research, then a year without a good idea is a wasted year. If you do research and teaching, then, no matter how the research has gone, you will have done something useful. Finally, most people, even mathematicians, have an extrovert side. "The smell of the greasepaint and the roar of the crowd" is irresistible even when translated into "the feel of the chalk and the scratch of the pen."[1]

I was a student at Cambridge in the 1960s. Some of my lecturers belonged to an older generation. Later we learnt that many of them had done exciting things during the war, but, at the time, they just seemed incredibly old and staid. On the whole, this older generation approached their lectures by producing as perfect a set of notes as possible and then writing them out on the blackboard. An extreme example of this system was given by J. C. Burkill whose course on measure theory was a word for word transcription of his book *The Lebesgue Integral*. In retrospect, I consider this book to be an excellent introduction to the subject, but I still do not think that the lectures added much to my understanding.

The younger lecturers reacted against this older style in a variety of ways. Instead of giving the same perfected course each year, they preferred three year stints on the principle that "the first year you learn, the second year you teach, the third year you embroider and the fourth year is worse." Some of them produced elaborate printed notes in order that the students could concentrate on the lectures without being distracted by the business of taking notes. So far as I was concerned, this was a failure, either because taking notes actually helps concentration, or because I found it hard to read notes and listen at the same time, or for some other reason. This experience must have been widespread since few lecturers of my generation produce such notes and many of those who do only give them to the students *after* the relevant lecture.

On the whole, the lecturers who left the greatest impression on me were those like Varopoulos, Conway and Swinnerton-Dyer both for their evident swagger (who can forget Swinnerton-Dyer arguing that "nature cannot be so unkind as to allow this result to be false") and the fact that they lectured without notes. Of course, "making an impression" is not quite the same as "giving a good course" or "teaching well," but it is an excellent start.[2]

Later, I spent the third year of my Ph.D. at Orsay (one of the Parisian universities). On arrival, I was told that I had been put down to give a seminar in four weeks' time on a paper of Kahane. Naturally, the seminar would be in French and without notes. Since I did not know Kahane's paper and my French was limited to four years in school and a few further hours in language labs, the four weeks were rather nerve wracking but, so far as I was concerned, the seminar went fine[3] (the audience may have felt differently) and the thrill I experienced made me a convert to noteless lecturing.

> Watching a lecture by Bolobás or a similar master of the art of lecturing without notes is like watching a magician give a stage performance. The lecturer may not be naked but, at least, strips down to the undergarments to show that nothing is hidden from the audience. There is a roll of drums and then "look ladies and

---

[1] There is claimed to be a survey in which 94 percent of professors said they were better teachers than the average faculty member on campus. I have been unable to track down such a survey but it sounds horribly plausible.

[2] It can be argued that the purpose of lecturing is not so much to teach as to provide the audience with a set of stories to tell their children.

[3] The secret to lecturing in a foreign language is to learn the first five minutes by heart and then rely on adrenaline.

gentlemen no artificial aids" a delicate inequality is conjured from thin air or some intractable theorem wrestled to the ground.

Why is this performance (for it is a performance), so valuable? I think there are two reasons. The first is the moral effect. Mathematics students faced with a new result have a natural tendency to believe that it is too hard for anybody to understand properly. If you copy out a proof on to the board or flash up the proof on a projector, the implicit message is that the proof is too hard for you to do anything but copy it out word for word. If you produce the proof without notes, the implicit message is that the proof is so easy that it is not worth making a fuss about.

There is a second reason for this style of lecturing. Mathematics is not a collection of facts but of processes. A slide show (and what is a computer presentation but a slide show?) of theorems and their proofs is like walking through a museum full of stuffed animals. Only by watching you actually proving the results can your audience see the animals live in their native habitats.

The non-mathematician may ask how a lecturer can possibly remember 50 minutes of mathematics. The answer is that she does not. Many proofs are entirely routine and can be constructed on the fly. Most of the remainder require one, or, at most, two, ideas and, once those are understood, the rest of the proof is again routine. A mathematics lecture is not like a classical symphony but like a jazz improvisation starting from a small number of themes.[4]

Boswell recorded Johnson's opinion that

> People have now-a-days got a strange opinion that everything should be taught by lectures. Now I cannot see that lectures can do so much as reading the books from which the lectures are taken. I know nothing that can be best taught by lectures except where experiments are shown. You may teach chemistry by lectures:—you might teach the making of shoes by lectures.

I believe that the making of proofs (as opposed to the discovery of new mathematics) is like the making of shoes. It is better taught by watching someone make a shoe in front of you than by trying to figure it out from books.

Even if the stage magician's act looks impromptu, it will be better if it is rehearsed. Varopoulos depends on his native wit and knowledge to carry him through his lecture. Since he has ample supplies of both, five lectures out of six would be splendid bravura performances and the sixth would misfire completely. Conway would ask for the syllabus five minutes before a lecture and then lecture brilliantly, but this reflected the fact that, over the years, he had reflected deeply on almost every topic to be met in the undergraduate course.

The rest of us need to prepare in advance. (Some lecturers like Beardon produce a beautiful set of notes and then lecture without consulting them, but few aspire to this standard.) When I started as a lecturer, I used to try out each lecture in an empty lecture room. Nowadays, I am too self conscious and lazy to do this (although I suspect it represents best practice) and, instead, I work through the next day's lecture the evening before. Only

---

[4] Unfortunately names are facts and not processes. If I need to use the name of a mathematician during a lecture I write it on my cuff or an equivalent surface.

the most self-confident mathematician will find the eigenvalues of a 3×3 matrix in public without having gone through the calculation privately several times before.

It is also vital to have an undisturbed ten minutes before the lecture. It does not matter if you do not think about the lecture. It does matter that you do not think about anything else. I have discussed this with colleagues and we all agree that, however well you prepare in advance, going directly from a committee meeting or similar occasion to give a lecture is a recipe for disaster.

A high wire act is not a real high wire act unless the performer might fall off. What happens when you fall off? The first lecture of my first course ran into such difficulties that I had to repeat the entire lecture next time and, the worst that can happen having happened, I no longer fear it.[5] The key advice for a lecturer who has got lost in an exposition is "when you find yourself in a hole stop digging." Tell the audience that you need to think, step back from the blackboard and reflect. If you see your way clear, return to the blackboard and continue. If you cannot see your way clear or you have failed in a second attempt, tell the audience what the problem is and that you will return to the proof in the next lecture. Some lecturers feel that it is important to give the correct proof at the blackboard (on the principle that, when you fall off a horse, you should immediately remount) but my experience is that, once things go wrong, they tend to continue that way and I prefer to write out the proof before the next lecture and hand out copies to the audience.

Mathematicians and those well on the way to being mathematicians understand that mathematics is difficult and that, from time to time, things will go wrong. The advice of the previous paragraph only applies to advanced courses. Beginning students are unforgiving. I once started a course with the words "I am not the world's greatest expert in this subject" and never fully recovered the confidence of the audience. "It is one of the first duties of a professor" writes Hardy "to exaggerate a little both the importance of his subject and his own importance in it." Fortunately, beginning students automatically assume that you are one of the world's greatest experts in your subject, so no lying is necessary. Since the mathematics taught to beginning students is necessarily elementary, it is highly unlikely that you will get lost but, if this happens, bluff ("I have given you the general idea, so go away and try and fill in the details. If you can't, I will give them next time") may be better than confession.

Failure at the blackboard may be nothing more that "blackboard blindness" but may indicate a gap in your own understanding. After many years of presenting the Hahn–Banach theorem smoothly and successfully, I gave a lecture in which I got totally confused. but, as a result, I feel I now understand the workings of the theorem much better.

What are the disadvantages of lectures given without notes? The first is that the students' notes will not be as perfect as if you copy or project previous written notes onto the board and the students copy those notes. There will be more small errors and proofs invented on the fly will not have the finished elegance of those written out beforehand. There is no doubt that weaker students value complete and accurate notes more than anything else. (The problem is that, having got complete and accurate notes, weaker students can do nothing with them.)

---

[5] The students for that course included Terry Lyons, Jonathan Partington, Keith Carne and Richard Pinch. For the rest of the course, whenever there was a gap in my reasoning one of them would point it out and another would explain how it should be filled. It was an exhilarating experience but such audiences are rather rare.

Those who agree with the weaker students that the main purpose of lectures is to produce complete and accurate notes must answer the traditional question "Did Gutenberg live in vain?" Unless we believe that a university education may be summarised as "take notes, learn notes, pass exam, forget notes" we should include among our educational objectives that students should learn to use libraries and consult books (or failing that, that they should consult Wikipedia with its many excellent mathematics articles).

Let me repeat what I said earlier. Mathematics is not a collection of facts but of processes. You cannot learn to ride a bicycle or play the violin from lectures. Instead you watch others riding bicycles or playing the violin and try to imitate them. You learn by long and painful practice (in the case of the bicycle painful to yourself, in the case of the violin painful to others). In the same way, you can only learn mathematics by doing exercises. It is possible (but quite hard) to learn mathematics without lectures by just reading books and doing exercises. It is possible (though, in my view, slightly unsatisfactory) to learn mathematics without books by just attending lectures and doing exercises. It is impossible to learn mathematics by just attending lectures and reading books.

I believe that lectures are only a small part of a mathematical education and that students learn most of their mathematics in other ways. I think that what students gain from lectures is the picture of a mathematician at work. By watching how she approaches a proof or how she always an example and a counterexample for each definition they are initiated into the mathematician's mode of thought. The cold perfection of books needs to be supplemented by the vision of mathematics as a living thing subject to human error.

The second disadvantage of lecturing without notes is organisational. Even if you get the class to help you number theorems consecutively, neither you nor your class will know the "the number of the theorem that we proved last Friday where we dropped the condition on differentiability." (Incidentally, lecturers should constantly bear in mind that the only person in the room who has been paying full attention all the time is the lecturer herself.) In addition you may suffer from notational drift by which the function f of the previous lecture becomes the function g of today. For this reason, I now mainly use a modified version of the noteless lecture in which I give the students a skeleton set of notes consisting of the statements of the definitions and theorems of the course and lecture following those notes.

2 Seminars

So far I have been concerned with lecture courses. What about seminars and similar occasions? The first thing to say is that the shorter the time given, the harder the task. If you lecture for 24 hours, you can build up a rapport with your audience. Ideally, you can move towards a more conversational ambience in which they do not hesitate to answer your questions or to ask their own. Even if this is not possible, you will be able to adjust your lecturing style (speed, sophistication, number and type of examples, ...) to match their reactions. If you have only sixty minutes, matters are much harder[6] and it may be impossible to rectify the choice of an inappropriate level. (I have been to very many seminars where

---

[6] I once attended an AMS meeting with 10 minute lectures. I think that the speakers did a splendid job, but I have no idea how they did it.

the speaker has pitched the level too high and very few where the speaker has pitched the level too low.)

When I discussed lecture courses, I made it clear that I considered them as a fairly minor part in education of our students. In the same way I think that seminars can only play a minor role in the education of their elders. In lecture courses we spend a long time trying to explain well understood things to our students with only limited success. In seminars we spend a short time trying to explain complicated and badly understood things to each other. Since, on the most optimistic view, we are only slightly more capable than our students it is foolish to expect too much.

What do I expect, or at least hope for, when I attend a seminar? In general I hope to understand the first few minutes, because the lecturer is supposed to be telling me things I already know. In general, I expect not to understand most of the rest, since the lecturer will be talking about things that I do not know and therefore cannot understand. But, during the few instants of changeover from one phase to the other, I hope to gain some insight into the subject of the seminar that I cannot gain from the books and papers bound to the formality of written mathematics. A good seminar talk will tell the audience what problem is being considered, where it comes from, why it is important and give some hint of what ideas are used to attack it.[7] If the speaker manages to get anything further across, the audience should consider this an uncovenanted bonus.

A *colloquium* is intended for an entire department. Most of the audience will not have studied your branch of mathematics since they were students. If you pitch your talk at the level of second year students that is probably the right level.

The non-mathematician may observe that, if my account of seminars is correct, then most of a seminar audience spends most of its time not understanding what the speaker is saying. It is a matter of observation that most non-mathematicians who fail to understand something in a talk blame the lecturer or, less usually, themselves.[8] Mathematicians are used to not understanding things and do not see that there is any blame to assign.

One of the leading British mathematicians once told me that he experienced a near breakdown in his third year of university. Up to then, he had understood everything in the lecture courses but this was no longer true. All his friends had gone through the same process much earlier, so he had to deal with this by himself. He argued that everyone had to deal with this crisis at some time and what mattered was not the timing of the crisis but how you cope with it.

Mathematicians are very good and polite audiences. One of my finest memories is of a conference dinner in Finland held in a farmhouse. At the end of the dinner, the local dramatic society performed a twenty minute skit in Finnish dealing with recent Finnish politics. The audience of Japanese, French and German mathematicians watched with rapt attention and applauded noisily at the end.

When I talked about lectures, I said that they were rarely useful for research. This is not true of seminars about one's own work. Writing a paper is a bit like assembling a piece of

---

[7] Some seminar speakers think that their job is to present a long list of theorems. Analytic number theorists seem particularly prone to the style "In 1953, X proved that the growth was no greater than $(\log x)^{2/3}$ but in 1957 Y improved this to $(\log x)^{20/31}(\log \log x)^{1/5}$." The Russian school enlivens the recital with details of priority disputes.

[8] This links with the belief common in the cultured classes that anything can be explained in five minutes.

complicated machinery. You have to concentrate on making sure that every part is free of defects and links correctly to its neighbours. Preparing a seminar forces you to take a step back and look at the piece of work as a whole. What are the principles behind the design of the machine? What obstacles have you overcome? What obstacles remain that prevent you from using the machine to overcome other problems?

If there is an expert on your topic in the audience, then you have the satisfaction of addressing her in person rather than through a paper which she may or may not read. She may find it much easier to grasp your underlying ideas in the context of an informal seminar than if she has to hack her way through the undergrowth of formal proof in your paper. Occasionally she may ask a question or make a remark that changes your ideas. (This does not happen very often. It has only happened to me on a handful of occasions and I have only understood the point being made sometime later. But those handful of occasions have played a major rôle in my research.)

Unfortunately, if there is an expert in the audience, it is very difficult to avoid addressing your talk to her and ignoring the rest of the audience. This is always rude and often unproductive since the expert may not be as expert as you think. (However low in the mathematical hierarchy you may feel yourself to be, you are probably the world expert on your own work.) Speaking from repeated experience, I know that this is a very difficult trap to avoid. I suggest that you adopt a rigid policy of addressing the general audience for, say, the first half of the seminar and only then take the presence of the expert or experts into account.

There is a another type of seminar that is so foreign to the Anglo-Saxon tradition that I know of it only through other people's accounts. This is announced as *X's Seminar*. Professor X chooses the speaker, subject and, in some cases, the audience. The speaker's job is to explain her topic to Professor X. The seminar speaker talks, with frequent interruptions from Professor X and his favourite pupil, until either Professor X is convinced that the whole thing is trivial or wrong, in which case he informs the speaker of the fact, or until Professor X has explained the topic to the lecturer. It was said that political meetings in the old Soviet Union resembled mathematics seminars but mathematics seminars resembled political meetings. It is clear that, if you have the right Professor X, the recipe produces a memorable and productive event. It is also clear that the right Professor X is a rather rare character.

Proportionally, I have given many more bad seminars than bad lecture courses. Partly this is due to the extra difficulties of the seminar form, but there is a further problem. A lecture course is a regular event and it is not hard to cultivate a regular habit of preparation. In contrast, having promised to give a talk in a couple of months, it is natural to forget about it or simply procrastinate until there is simply not enough time to prepare properly. Usually you can wing it successfully, but not always.

Mathematicians may also be asked to give "popular lectures." If the audience is under eighteen, there may well be an element of compulsion in their decision to attend. If the audience consists of adults, they have chosen to come and they are either convinced or wish strongly to be convinced that mathematics is interesting and useful. In either case your sole duty is to entertain. Speaking personally, such lectures lack the edge that comes from the possibility of disaster and the satisfaction of avoiding that disaster. It may (indeed it does) sound pompous but I cannot imagine any audience for a popular lecture that would give me the same thrill as I got from giving the first lecture to beginning undergraduates at Cambridge, knowing that, in all likelihood, the audience contained a student whose name would be remembered when mine is forgotten.

There is a problem with general talks attributable to the spirit of the age. Some years ago, I attended a conference for mathematicians and engineers. The mathematicians felt insulted by the engineers' flashy computer presentations which were clearly used and reused for many occasions. The engineers felt insulted by the blackboard lectures of the mathematicians which were clearly just thrown together for that particular conference. PowerPoint and its relatives make it possible for people to talk in public who previously lacked the confidence to do so.[9] Unfortunately this means that many audiences (not, however, mathematicians) expect a PowerPoint presentation and are disappointed when they do not get it.

No mathematician who reads this will be surprised by my recommendation (valid for the year 2015), that if a computer presentation is required, they should use Beamer. There are three reasons in increasing order of importance. First Beamer is "open source" and, because mathematics as practised for the last 3000 years has been "open source," mathematicians are ideologically attracted to such programs. Second, PowerPoint is not designed to communicate mathematics (or, indeed, according to Edward Tufte, to communicate anything at all). Third, and most importantly, Beamer is a $\LaTeX$ system and $\LaTeX$ is now the language of mathematical printing.

Finally, a small book could be written about the problems of lecturing in other people's lecture rooms, starting with those rooms designed by an architect who once met a man who had an aunt who had actually attended a lecture.[10] Among the many things that may not be provided are coloured chalks, white chalk that does not crumble in your hand, any chalk at all, whiteboard markers that actually mark the board, whiteboard erasers, overhead projectors, transparencies for overhead projectors, working pens for overhead transparencies, computer projection systems compatible with your computer and computers that can understand your memory stick. It is also a rule that all computer systems will fail mysteriously ten minutes into any talk. If you are prepared for such problems, they may not materialise. If you are unprepared, they almost certainly will.[11] Make sure that you see the lecture room well in advance and check all the equipment you use. Insist that your host's unlimited hospitality includes leaving you alone for ten minutes before the talk to collect your thoughts.

3 Reflections

My discussion has probably annoyed non-mathematicians, mathematicians and educationalists. However it has annoyed them in different ways.

---

[9] Whether the sum of human happiness is increased thereby is another matter. The pleasure of the PowerPoint presenter is often bought at the expense of the audience.

[10] The following advice is intended for any architect who reads this. Remember the reason lecture theatres are so called is that they are *theatres* with the lecturer as the star. Look at the lecture theatre at the Royal Institution in London to see how such a theatre should be built.

[11] A historian friend of mine was told that the room for her talk was fully equipped for computer presentations. When she arrived with her memory stick it turned out that this meant it had a portable projection-screen.

To the non-mathematician the tone of the discussion probably appears insufferably arrogant. But any working mathematician must have a touch of arrogance. If you tackle an unsolved problem you must believe that you can succeed where others have failed. You may think this is because you are cleverer than they were, or because you know more, or because you have a new approach or just because you are prepared to work harder and longer (it is quite possible for a problem to be quite well known but never to have been seriously attacked). Note that this arrogance is of a limited nature. There is a very old story of two men being chased by a bear. One of them shouts "Its no good, you can't outrun the bear" and the other shouts back "I'm not trying to outrun the *bear*." When you try to solve a problem you do not have to outrun all the people who *could* have attacked the problem, only those that have.

Lecturing is very different from research but unless you believe that you know things that your audience does not know, that those things are worth knowing and that you can teach them to your audience, it is not clear why you should be lecturing to them. Modesty is out of place in front of the blackboard.

Professor M was not only one of the deepest mathematicians in the world but also one of the most modest. When he lectured he apologized to the audience for the triviality of what he was going to present. In his anxiety not to bore the audience he lectured faster and faster with more and more apologies. It is said that, when he went on a lecture tour of U.S. universities, he was followed by another lecturer who gave a second lecture to explain Professor M's original lecture.

One of the many paradoxical elements in teaching is that while students who accept nothing that you say will do badly, students who accept everything you say will not do particularly well. However, since good students mainly educate themselves outside the lecture room, you need not worry that you will influence them too much.[12] Eventually they will outgrow you. The clever ones will make this clear to you but the very clever ones will listen to you as respectfully as ever.

Mathematicians will be annoyed by the fact that I have not dealt with the problem of lecturing to those unwilling to learn. This reflects the fact that (as they have been muttering through clenched teeth) "The lines have fallen to me in pleasant places." It also reflects the fact that I have very little advice to give. In the rare cases when your audience is willing to learn from others but not from you, it may be worth asking yourself why you are teaching them what you are teaching them. If it is going to be useful to them, take time out to explain to them why it is going to be useful to them. If it is not, then perhaps you should be teaching them something else.

Educationalists will be annoyed that I have not made any reference to their labours. Like most mathematicians, I remain unconvinced that they have, as yet, much to offer. They are welcome to put this down to arrogance (see above) and intellectual laziness but they would do better to reflect on the reasons we give for our views.

(1) Much of what is quoted as educational research is mere expression of opinion. Much of the rest hardly rises above what Feynman calls cargo cult science (see *Surely You're Joking Mr Feynman!* again).

---

[12] And, if you do influence them, it will be in unpredictable ways. One of my former students told me that he "... will never forget the way you told us to think, don't calculate" while another remembered the way I told him that "... calculation is the way to truth."

(2) The only outcomes of education that we can measure (student satisfaction, recall of material after two months and so on) are unsatisfactory proxies for whatever outcome it is that we actually wish for.

(3) Education is not a generic process. The education required to produce a literary critic, a chemist, a violinist or a mathematician is very different. It is now customary in British Universities to provide courses on lecturing to new lecturers.[13] Often such courses include the advice that you should only put three sentences on a blackboard at at a time. So far as mathematicians are concerned, one might as well advise mountaineers to avoid steep slopes or surgeons only to operate on the exceptionally healthy.

(4) Many good mathematicians are also good lecturers. (It is only youthful innocence that causes our students to believe that those who lecture badly do so because they are great thinkers.) After all, good research usually demands insight and clarity and these virtues provide a very good foundation for good lecturing. However, if it is necessary to make a choice, most mathematicians would prefer someone with something to say but who says it badly to some one with nothing to say who says it brilliantly. The better mathematician trumps the better lecturer.

Having said all this, I do believe that a lecturer who thinks about the process of lecturing will lecture better than one who does not. By taking thought, a dreadful lecturer can become a bad lecturer, a bad lecturer can become a mediocre lecturer, a mediocre lecturer a good lecturer, and a good lecturer an excellent one. If you go to a lecture or seminar that you enjoy you should ask yourself why it went so well. If you go to a lecture or seminar which you disliked, ask yourself what the speaker did wrong. If you give a lecture that goes particularly well or particularly badly ask yourself why. (Often the answer is obvious and unhelpful, but not always.)

There are some simple pieces of advice.[14] Look at the clock from time to time. Try not to end in the middle of a proof. (You will only have to start again at the beginning next time. Either finish early or waffle a bit about what is coming.) Start each lecture by summarising where you have got to. Finish each lecture by saying what you have done. Pause from time to time to allow the audience to catch up on its note taking and to catch its intellectual breath. You can produce such a pause either by stopping speaking (very hard, if you decide to be silent for 30 seconds, you will need to time yourself by the watch, otherwise you will restart after 10 seconds) or by talking about non-essential matters like the history of the problem or the traffic on the way to work.[15] After lectures, walk to the other end of the room to check that your blackboard writing is legible.[16]

It is not entirely surprising that our custom of giving students a morning of advice on attending lectures before they have actually attended one, sometimes fails in its intended purpose. I suspect that the advice would be more effective if it was given after two weeks of lectures, when the students actually know what a lecture is. In the same way, I doubt

---

[13] My younger colleagues suggest that I should include advice to those who organise such courses that, other things being equal, it might, perhaps, be better if courses on lecturing are given by people who can lecture.

[14] Complicated pieces of advice are no use. A swimming trainer once told me that it took 24 hours of practice to change any aspect of a swimmer's style. There is just too much going on in a lecture to allow you to follow anything but the simplest advice.

[15] Dr R endeared himself to generations of Cambridge students by complaining about his dreadful hangovers.

[16] I once asked Conway the best way to finish a lecture course. He replied "Early."

if lecturing training is very effective if it is given to those who have not lectured before. If I were a vice-chancellor [17] I would substitute occasional meetings with plenty of cream cakes[18] to allow lecturers to talk about problems and exchange hints of the type given in the previous paragraph.

Those who see mathematicians as irredeemably conservative should reflect that university mathematical education does change with time and the path to the modern mathematics departments runs from the École Polytechnique in 1800 through Göttingen in 1910. They should also reflect that the system is an almost ideally Darwinian one. The most successful teachers will have the most successful students and they will teach, with substantial variation, in the way they were taught. If a consistently better teaching method appears it should spread rapidly.

Mathematicians have their heroes. When I was a young mathematician at Orsay, the word went round that Gelfand had been allowed out by the Soviet authorities to visit Paris and would give a lecture at 4 PM. The lecture room was filled to capacity and beyond by mathematicians from all over Paris with the most eminent in the front row[19] and late arrivals sitting on the steps or standing at the back. There was a storm of applause and Gelfand (yes, the Gelfand whose name figured so prominently in my fourth year studies, yes, Gelfand himself) arrived visibly tired from his long journey but visibly delighted to be where he was, shaking hands enthusiastically with the entire front row. The lecture was beautifully delivered in excellent English but at a level of abstraction well beyond me. All I remember is an elegant formula suddenly appearing and Gelfand saying that "A concrete formula like this reassures us that we must be on the right track." Those who think that the only purpose of lectures is to communicate knowledge will not understand why Gelfand's lecture is a treasured memory.

---

[17] "Which, thank the lord, I'm not, sir."
[18] The cream cakes are particularly important.
[19] When I was young, I thought that the old sat in the front to demonstrate their eminence. Now that I am older I realise their choice has more to do with the state of their eyes and ears.

7
Through a Glass Darkly

Steven G. Krantz[1]

1 Prolegomena

Education is a repetition of civilization in little.

—Herbert Spencer

Being a research mathematician is like being a manic depressive. One experiences occasional moments of giddy elation, interwoven with protracted periods of black despair. Yet this is the life path that we choose for ourselves. And we wonder why nobody understands us.

The budding mathematician invests an extraordinarily long period of study—four years of undergraduate study and five years or more in graduate school—in order to attain the Ph.D. And that is only an entry card into the profession. It hardly makes one a mathematician.

To be able to call yourself a research mathematician, you must have proved some good theorems and written some good papers thereon. You must have given a number of talks on the work, and (ideally) you should have either an academic job or a job in the research infrastructure. Then, and only then, can you hold your head up in the community and call yourself a peer of the realm. Often you are thirty years old or older before this comes about. It is a protracted period of apprenticeship, and there are many souls fallen and discouraged and indeed lost along the way.

The professional mathematician lives his life thinking about problems that he/she cannot solve, and learning from his/her repeated and often maddening mistakes. That he/she can very occasionally pull the fat out of the fire and make something worthwhile of it is in fact a small miracle. And even when he/she can pull off such a feat, what are the chances that his/her peers in the community will toss their hats in the air and proclaim him/her a hail fellow well met? Slim to none at best.

---

[1] It is a pleasure to thank Don Albers, David H. Bailey, Jonathan Borwein, Robert Burckel, David Collins, Marvin Greenberg, Reece Harris, Randi Ruden, and James S. Walker for many useful remarks and suggestions about different drafts of this essay. Certainly their insights have contributed a number of significant improvements.

In the end, we learn to do mathematics because of its intrinsic beauty, and its enduring value, and for the personal satisfaction it gives us. It is an important, worthwhile, dignified way to spend one's time, and it beats almost any other avocation that I can think of. But it has its frustrations.

There are few outside of the mathematical community who have even the vaguest notion of what we do, or how we spend our time. Surely they have no sense of what a theorem is, or how to prove a theorem, or why anyone would want to.[2] How could you spend a year or two studying other people's work, only so that you can spend yet several more years to develop your own work? Were it not for tenure, how would any mathematics ever get done?

We in the mathematics community expect (as we should) the state legislature to provide funds for the universities (to pay our salaries, for instance). We expect the members of Congress to allocate funds for the National Science Foundation and other agencies to support our research. We expect the White House Science Advisor to speak well of academics, and of mathematicians in particular, so that we can live our lives and enjoy the fruits of our labors. But what do these people know of our values and our goals? How can we hope that, when they do the obvious and necessary ranking of priorities that must be a part of their jobs, we will somehow get sorted near the top of the list?

Our frequent relegation to the bottom of that list explains in part why we as a profession can be aggravated and demoralized, and why we endure periods of frustration and hopelessness. We are not by nature articulate—especially at presenting our case to those who do not speak our language—and we pay a price for that linguistic lassitude. We tend to be solipsistic and focused on our scientific activities, and trust that the value of our results will speak for themselves. When competing with the Wii and the iPod, we are bound therefore to be daunted.

2 Life in the Big City

> The most savage controversies are about those matters as to which there is no good evidence either way.
>
> —Bertrand Russell

If you have ever been chair of your department, put in the position of explaining to the dean what the department's needs are, then you know how hard it is to explain our mission to the great unwashed. You waltz into the dean's office and start telling him how we must have someone in Ricci flows, we certainly need a worker in mirror symmetry, and what about that hot new stuff about the distribution of primes using additive combinatorics? The dean, probably a chemist, has no idea what you are talking about.

To make matters worse, the person with the previous appointment with the dean was the chair of chemistry, and he glibly told the dean how they are woefully shy of people in radiochemistry and organic chemistry. And that an extra physical chemist or two would be nice as well. The dean said "sure," he understood immediately. It was a real shift of gears then for the dean to have to figure out what in the world you (from the mathematics

---

[2] From my solipsistic perspective as a mathematician, this is truly tragic. For mathematical thinking is at the very basis of human thought. It is the key to an examined life.

department) are talking about. How do you put your case into words that the dean will understand? How do you sell yourself (and your department) to him?[3]

Certainly we have the same problem with society at large. People understand, just because of their social milieu, why medicine is important and useful. Computers and their offspring make good sense; we all encounter computers every day and have at least a heuristic sense of what they are good for. Even certain parts of engineering resonate with the average citizen (aeronautics, biomedical engineering, civil engineering). But, after getting out of school, most people have little or no use for mathematics. Most financial transactions are handled by machines. Most of us bring our taxes to professionals for preparation. Most of us farm out construction projects around the house to contractors. If any mathematics, or even arithmetic, is required in the workplace it is probably handled by software.

One of my wife's uncles, a farmer, once said to me that we obviously no longer need mathematicians because we have computers. I gave him a patient look and said yes, and we obviously no longer need farmers because we have vending machines. He was not amused. But the analogy is a good one. Computers are great for manipulating data, but not for thinking. Vending machines are great for handing you a morsel of food *that someone else has produced in the traditional fashion*.

People had a hard time understanding what Picasso's art was about—or even Andy Warhol's art—but they had a visceral sense that it was interesting and important. The fact that people would spend millions of dollars for the paintings gave the activity a certain gravitas, but there is something in the nature of art that makes it resonate with our collective unconscious. With mathematics, people spend their lives coming to grips with what was likely a negative experience in school, reinforced by uninspiring teachers and dreadful textbooks. If you are at a cocktail party and announce that you don't like art, or don't like music, people are liable to conclude that you are some sort of philistine. If instead you announce that you don't like mathematics, people will conclude that you are a regular guy. (If you choose to announce that you *do* like mathematics, people are liable to get up and walk away.) To the uninitiated, mathematics is cold and austere and unforgiving. It is difficult to get even an intuitive sense of what the typical mathematician is up to. Unlike physicists and biologists (who have been successfully communicating with the press and the public for more than fifty years), we are not good at telling half-truths that paint a picture of our meaning and get our point across. We are too wedded to the mathematical method. We think in terms of definitions and axioms and theorems.

3 Living the Good Life

One normally thinks that everything that is true is true for a reason. I've found mathematical truths that are true for no reason at all. These mathematical truths are beyond the power of mathematical reasoning because they are accidental and random.

—G. J. Chaitin

---

[3] It is arguable that a mathematics department is better off with a dean who is a musicologist or perhaps a philologist. Such a scholar is not hampered by the *Realpolitik* of lab science dynamics, and can perhaps think imaginatively about what our goals are.

The life of a research mathematician is a wonderful experience, an exhilarating, blissful existence for those who are prone to enjoy it. You get to spend your time with like-minded people who are in pursuit of a holy grail that is part of an important and valuable larger picture that we are all bound to. You get to travel, and spend time with friends all over the world, and hang out in hotels, and eat exotic foods, and drink lovely drinks. You get to teach bright students and engage in the marketplace of ideas, and actually to develop new ones. What could be better? There is hardly a more rewarding way to be professionally engaged.

It is a special privilege to be able to spend our time—and be paid for it—thinking original (and occasionally profound) thoughts and developing new programs and ideas. We actually feel that we are changing the fabric of the cosmos, helping people to see things that they have not seen before, affecting people's lives.[4] Teaching can and probably should be a part of this process. For surely bringing along the next generation, training a new rank of scholars, is one of the more enlightened and certainly important mathematical pursuits. Also interacting with young minds is a beautiful way to stay vibrant and plugged in, and to keep in touch with the development of new ideas.

Of course there are different types of teaching. The teaching of rudimentary calculus to freshmen has different rewards from teaching our latest research ideas to graduate students. But both are important, and both yield palpable results. What is more, *teaching is an activity that others understand and appreciate*. If the public does not think of us as air-headed scholars or spoiled elitists, then surely it occasionally thinks of us as teachers. And better that *we* should have to do that teaching. After all, it is our bailiwick.

The hard fact of the matter is that the powers that be in the university also appreciate our teaching rather more than they do our many other activities. After all, mathematics is a key part of the core curriculum. A university could hardly survive without mathematics. Other majors could not function, could not advance their students, could not build their curricula, without a basis in mathematics. So our teaching role at the institution is both fundamental and essential.[5] Our research role is less well understood, especially because *we do not by instinct communicate naturally with scholars in other departments*.

This is actually a key point. We all recall the crisis at the University of Rochester twenty years ago, when the dean shut down the graduate program in mathematics. His reasoning, quite simply, was that the mathematics department was isolated, did not interact productively with other units on campus, did not carry its own weight. The event at Rochester tolled a knell throughout the profession, for we all knew that similar allegations could be leveled at any of us. Institutions like Princeton or Harvard are truly ivory towers, and unlikely to suffer the sort of indignity being described here. But if you work at a public institution, then look out. I work at a *very* private university, and I can tell you that, in my negotiations as chair with our dean, he sometimes brought up Rochester. And he did *not* do so in an effort to be friendly. He was in fact threatening me.

Some departments, like earth & planetary science or biomedical engineering, interact very naturally with other subjects. Their material is intrinsically interdisciplinary. It makes perfect sense for these people to develop cross-cultural curricula and joint majors with

---

[4] I have long been inspired by Freeman Dyson's book [DYS]. It describes both poignantly and passionately the life of scientists, and how they can feel that they are altering and influencing the world around them.

[5] This fact is recognized in promotion and tenure proceedings. The criteria for promotion have always been teaching, research, and service. In recent years, teaching has played a more dominant role. And it should.

other departments. It is very obvious and sensible for them to apply for grants with people from departments even outside of their school. A faculty member in such a department will speak several languages semi-fluently.

It is different for mathematics. It is a challenge just to speak the one language of mathematics, and to speak it well. Most of us do a pretty good job at it, and those outside of mathematics cannot do it at all. So there is a natural barrier to communication and collaboration. In meetings with other faculty—even from physics and engineering—we find it difficult to identify a common vocabulary. We find that we have widely disparate goals, and very different means of achieving them.

Also our value systems are different. Our methods for gauging success vary dramatically. Our reward systems deviate markedly. Different departments use quite different criteria for promotion and tenure. Some emphasize the ability to interact with colleagues while others will emphasize lone research efforts. Some emphasize teaching while others will emphasize writing. Once you become a full professor you serve on tenure and promotion committees for other departments. This experience is a real eye-opener, for you will find that the criteria used in English and history and geography are quite different from what we are accustomed to.[6] Even our views of truth can be markedly different.

4 The Why and the Wherefore

> The lofty light of the a priori outshines the dim light of the world and makes for us incontrovertible truths because of their "clearness and distinctness."
>
> —René Descartes

A mathematician typically goes through most of his/her early life as a flaming success at everything he/she does. You excel in grade school, you excel in high school, you excel in college. Even in graduate school you can do quite well if you are willing to put forth the effort.

Put in slightly different terms: You can get a long way in the basic material just by being smart. Not so much effort or discipline is required. And this may explain why so many truly brilliant people get left in the dust. They reach a point where some real *Sitzfleisch* and true effort are required, and they are simply not up to it. They have never had to expend such disciplined study before, so why start now?

While there is no question that being smart can take you a long way, there comes a point—*for all of us*—where it becomes clear that a capacity for hard work can really make a difference. Many professional mathematicians put in *at least* ten hours per day, *at least* six days per week. There are many who do much more. And we tend to enjoy it. The great thing about mathematics is that it does not fight you. Generally speaking, it will not sneak behind your back and bite you. It tends to be both satisfying and rewarding.

Doing mathematics is *not* like laying bricks or mowing the grass. The quantity of end product is not a linear function of the time expended. Far from it. As Charles Fefferman,

---

[6] I still recall serving on the committee for promotion to professor of a candidate in geography. One of his published writings was called *A Walk Through Chinatown*. It described the experience of walking down Grant Avenue in San Francisco and smelling the wonton soup. What would be the analogue of this in a case for promotion in mathematics?

Fields Medalist, once said, a good mathematician throws 90% of his work in the trash (and this percentage goes up as time goes on). Of course you learn from all that work, and it makes you stronger for the next sortie. But you often, at the end of six months or a year, do not have much to show.

On the other hand, you can be blessed with extraordinary periods of productivity. The accumulated skills and insights of many years of study suddenly begin to pay off, and you find that you have plenty to say. And it is *quite* worthwhile. Certainly worth writing up and sharing with others and publishing. This is what makes life rewarding, and this is what we live for.

Economists like to use the professoriate as a model, because it runs contrary to many of the truisms of elementary economic theory. For example, if you pay a Professor of Mathematics twice as much, that does not mean that he/she will be able to prove twice as many theorems, or produce twice as many graduate students. The truth is that he/she is probably already working to his/her capacity. There are only so many hours in the day. What more can he/she do? It is difficult to say what or how a professor of mathematics should be compensated, because we do not fit the classical economic model.

Conversely, we could also note that if you give a professor of mathematics twice as much to do, it does not follow that he/she will have a nervous breakdown, or quit, or go into open rebellion. Many of us now have a teaching load of two courses per semester. But sixty years ago the norm—even at the very best universities in the United States—was three courses (or more!) per semester. Also, in those days, there was very little secretarial help. Professors did a lot of the drudgery themselves. There were also no NSF grants, and very little discretionary departmental money, so travel was often covered from one's own pocket. Today life is much better for everyone.

It is getting increasingly difficult to obtain NSF grants in mathematics. Back in the early 1960s, any reasonable mathematician doing any reasonable thing had a grant. I recall a seasoned mathematician, around 1969, lamenting that "now the NSF expects me to write a paper in order for me to get my grant renewed." Not too many years later an NSF program officer said to me that, "Now you have to write one paper per year." Well, things have gotten much tougher. Today the grant holders are a rather small minority, and there are a number of outstanding researchers with no grant support. The reasons for this are several:

- The slice of the NSF pie for mathematics is rather small compared to that for the other sciences.
- As the David Report taught us, mathematics is easily confused with statistics and computer science. As a result, our money is often co-opted.
- Much NSF math money is siphoned off to support bricks-and-mortar institutes, to subsidize fellowships, to subvene activities like the summer meetings at Park City, and for many other non-research-subsidy activities.

Several years ago my university had a dean—a physicist—who was in office for about eight years. This dean's job was a huge one—really more like a provostship. He essentially ran the university. For eight years he certainly had no time to do research. But, as soon as he left office, he went straight back to the physics department and got a huge grant to do high-energy physics. And that is what he is doing these days.

What I have just described would be literally impossible in mathematics. In math, if you are out of the loop for eight years then you are virtually at the level of a beginning postdoc. You certainly cannot submit a competitive grant application, and it would a tough, life-and-death battle to get back into the research game.

This just shows how different mathematics is from other subjects. Just by the very nature of our infrastructure, and the competitive form of the game, we are more isolated than other scholars. And we are easily marginalized. But there are also good features of being a mathematician.

The fact is that a Professor of Mathematics has a good deal of slack built into his/her schedule. If you double his/her teaching load, it means that he/she has less time to go to seminars, or to talk to his/her colleagues, or just to sit and think. But he/she will still get through the day. Just with considerably less enthusiasm. And notably less creativity. Universities are holding faculty much more accountable for their time these days. Total Quality Management is one of many insidious ideas from the business world that is starting to get a grip at our institutions of higher learning. In twenty years we may find that we are much more like teachers (in the way that we spend our time) and much less like scholars.

Sad to say, the dean or the provost has only the vaguest sense of what our scholarly activities are. When they think of the math department at all, they think of us as "those guys who teach calculus." They certainly *do not* think of us as "those guys who proved the Bieberbach conjecture." Such a statement would have little meaning for the typical university administrator. Of course they are pleased when the faculty garners kudos and awards, but the awards that Louis de Branges received for his achievement[7] were fairly low key.[8] They probably would not even raise an eyebrow among the Board of Trustees.

5 Such is Life

> There is no religious denomination in which the misuse of metaphysical expressions has been responsible for so much sin as it has in mathematics.
> —Ludwig Wittgenstein

Mathematicians are very much like oboe players. They do something quite difficult that nobody else understands. That is fine, but it comes with a price.

We take it for granted that we work in a rarified stratum of the universe that nobody else will apprehend or appreciate. We do not expect to be able to communicate with others. When we meet someone at a cocktail party and say, "I am a mathematician," we expect to be snubbed, or perhaps greeted with a witty rejoinder such as, "I was never any good in math." Or, "I was good at math until we got to that stuff with the letters—like algebra."

When I meet a brain surgeon I never say, "I was never any good at brain surgery. Those lobotomies always got me down." When I meet a proctologist, I am never tempted to say, "I

---

[7] He was a plenary speaker at the International Congress of Mathematicians, conferences were held in his honor, and he got a big NSF Grant.

[8] When I was chair of the mathematics department, the dean was constantly reminding me that he thought of us as a gang of incompetent, fairly uncooperative boobs. One of his very favorite chairs at that time was the head of earth and planetary sciences. This man was in fact the leader of the Mars space probe team, and he actually designed the vehicle that was being used to explore Mars. Well, you can imagine the kind of presentations that this guy could give—lots of animated graphics, lots of panoramic vistas, lots of dreamy speculation, lots of stories about other-worldly adventures. His talks were given in the biggest auditoriums on campus, and they were always packed. The dean was front and center, with his tongue hanging out, every time; he fairly glowed in the dark because he was so pleased and excited. How can a mathematician compete with that sort of showmanship? Even if I were to prove the Riemann Hypothesis, it would pale by comparison.

was never any good at...." Why do we mathematicians elicit such foolish behavior from people?

One friend of mine suggested that what people are really saying to us, when they make a statement of the sort just indicated, is that they spent their college years screwing around. They never buckled down and studied anything serious. So now they are apologizing for it. This is perhaps too simplistic. For taxi drivers say these foolish things too. And so do mailmen and butchers. Perhaps what people are telling us is that they *know* that they should understand and appreciate mathematics, but they do not. So instead they are resentful or intimidated.

There is a real disconnect when it comes to mathematics. Most people, by the time that they get to college, have had enough mathematics to be pretty sure they do not like it. They certainly do not want to major in the subject, and their preference is to avoid it as much as possible. Unfortunately, for many of these folks, their major may require a nontrivial amount of math (not so much because the subject area actually *uses* mathematics, but rather because the people who run their department seem to want to use mathematics as a *filter*). And also unfortunately it happens, much more often than it should, that people end up changing their majors (from engineering to psychology or from physics to media studies) simply because they cannot hack the math.

In recent years I have been collaborating with plastic surgeons, and I find that this is a wonderful device for cutting through the sort of conversational impasse that we have been describing. *Everyone*, at least everyone past a certain age, is quite interested in plastic surgery. People want to understand it, they want to know what it entails, they want to know what are the guarantees of success. When they learn that there are connections between plastic surgery and mathematics, then that is a hint of a human side of math. It gives me an entree that I never enjoyed in the past.

I also once wrote a paper with a picture of the space shuttle in it. That did not prove to be quite so salubrious for casual conversation; after all, engineering piled on top of mathematics does not make the mathematics any more palatable. But at least it was an indication that I could speak several tongues.

And that is certainly a point worth pondering if we want to fit into a social milieu. Speaking many tongues is a distinct advantage, and gives you a wedge for establishing real communication with people. It provides another way of looking at things, a new point of contact. Trying to talk to people *about mathematics, in the language of mathematics, using the logic of mathematics* is not going to get you very far. It will not work with newspaper reporters and it also will not work with ordinary folks that you are going to meet in the course of your life.

6 Mathematics and Art

It takes a long time to understand nothing.

—Edward Dahlberg

Even in the times of ancient Greece there was an understanding that mathematics and art were related. Both disciplines entail symmetry, order, perspective, and intricate relationships among the components. The golden mean is but one of many artifacts of this putative symbiosis. Ancient discussions of design and proportion are another.

M. C. Escher spent a good deal of time at the Moorish castle the Alhambra, studying the very mathematical artwork displayed there. His insights served to fuel his later studies (which are considered to be a very remarkable synthesis of mathematics and art).

Today there is more pronounced recognition of the interrelationship of mathematics and art. No less an eminence than Louis Vuitton offers a substantial prize each year for innovative work on the interface of mathematics and art. Benoit Mandelbrot has received this prize (for his work on fractals—see [MAN]), and so has David Hoffman for his work with Jim Hoffman and Bill Meeks on embedded minimal surfaces (see [HOF]).

Mathematics and art make a wonderful and fecund pairing for, as we have discussed here, mathematics is perceived in general to be austere, unforgiving, cold, and perhaps even lifeless. By contrast, art is warm, human, inspiring, even divine. If I had to give an after-dinner talk about what I do, I would not get very far trying to discuss the automorphism groups of pseudoconvex domains. I would probably have much better luck discussing the mathematics in the art of M. C. Escher, or the art that led to the mathematical work of Celso Costa on minimal surfaces.

Of course we as mathematicians perceive our craft to be an art form. Those among us who can see—and actually prove!—profound new theorems are held in the greatest reverence (at least by their colleagues), much as artists. We see the process of divining a new result and then determining how to verify it much like the process of eking out a new artwork. It would be in our best interest to convey this view of what we do to the world at large. Whatever the merits of fractal geometry may be, Benoit Mandelbrot has done a wonderful job of conveying both the art and the excitement of mathematics to the public. Fields Medalist Michael Atiyah has a lovely presentation [ATI] on Beauty in Mathematics.

Those who wish to do so may seek mathematics exhibited in art throughout the ages. Examples are

- A marble mosaic featuring the small stellated dodecahedron, attributed to Paolo Uccello, in the floor of the San Marco Basilica in Venice.
- Leonardo da Vinci's outstanding diagrams of regular polyhedra drawn as illustrations for Luca Pacioli's book *The Divine Proportion*.
- A glass rhombicuboctahedron in Jacopo de' Barbari's portrait of Pacioli, painted in 1495.
- A truncated polyhedron, and various other mathematical objects, which feature in Albrecht Dürer's engraving *Melancholia I*.
- Salvador Dalí's painting *The Last Supper* in which Christ and his disciples are pictured inside a giant dodecahedron.

Sculptor Helaman Ferguson [FER] has made sculptures of a wide range of complex surfaces and other topological objects. His work is motivated specifically by the desire to create visual representations of mathematical ideas. There are many artists today who conceive of themselves, and indeed advertise themselves, as mathematical artists. There are probably rather fewer mathematicians who conceive of themselves as artistic mathematicians.

Mathematics and music have a longstanding and deeply developed relationship. Abstract algebra and number theory can be used to understand musical structure. There is even a well-defined subject of musical set theory (although it is used primarily to describe atonal pieces). Pythagorean tuning is based on the perfect consonances. Many mathematicians are musicians, and take great comfort and joy from musical pastimes. Music can be an opportunity for mathematicians to interact meaningfully with a broad cross section of our world. Mathematicians Noam Elkies and Robert E. Greene have developed wonderful presentations—even full courses—about the symbiosis between mathematics and music.

Mathematics can learn a lot from art, especially from the way that art reaches out to humanity. Part of art is the interface between the artist and the observer. Mathematics is like that too, but typically the observer is another mathematician. We would do well, as a profession, to think about how to expand our pool of observers.

7 Mathematics Versus Physics

> I do still believe that rigor is a relative notion, not an absolute one. It depends on the background readers have and are expected to use in their judgment.
> —René Thom

Certainly "versus" is the wrong word here. Ever since the time of Isaac Newton, mathematics and physics have been closely allied. After all, Isaac Newton virtually invented physics as we know it today. And mathematics in his day was a free-for-all. So the field was open for Newton to create any synthesis that he chose.

But mathematics and physics are separated by a common goal, which is to understand the world around us. Physicists perceive that "world" by observing and recording and thinking. Mathematicians perceive that "world" by looking within themselves (but see the next section on Platonism vs. Kantianism).

And thus arises a difference in styles. The physicist thinks of himself as an observer, and is often content to describe what he sees. The mathematician is *never* so content. Even when he "sees" with utmost clarity, the mathematician wants to confirm that vision with a proof—deriving results from first principles, or axioms. This fact makes us precise and austere and exacting, but it also sets us apart and makes us mysterious and difficult to deal with.

In 1989, Stanley Pons and Martin Fleischmann announced that they had produced, in effect, cold fusion in their laboratory. Of course this was very exciting. They decided to bypass the usual funding agencies (the National Science Foundation and the Department of Energy, for instance) and go directly to Congress for funding for their further research. After all, this discovery had considerable potential impact on the quality of life in the United States. The good news is that this action brought their "discovery" a great deal of attention. The bad news is that it turned out to be incorrect. Experts later said that they were the victims of "dirty testtubes." It is difficult to imagine what would be the analogy of this scenario in the context of pure mathematics. What mathematical discovery would cause one to go directly to Congress? And, if you did so, how could you hope to get them to understand what you are talking about? These are truly different worlds.

I once heard Fields Medalist Charles Fefferman give a lecture (to a mixed audience of mathematicians and physicists) about the existence of matter. In those days Fefferman's goal was to prove the existence of matter from first principles—in an axiomatic fashion. I thought that this was a fascinating quest, and I think that some of the other mathematicians in the audience agreed with me. But at some point during the talk a frustrated physicist jumped up and shouted, "Why do you need to do this? All you have to do is look out the window to see that matter exists!"

Isn't it wonderful? Different people have different value systems and different ways to view the very same scientific facts. If there is a schism between the way that mathematicians view themselves and the way that *physicists* see us, then there is little surprise that there is such a schism between our view of ourselves and the way that non-scientists see us. Most laymen are content to accept the world phenomenologically—it is what it is. Certainly it is not the average person's job to try to dope out why things are the way they are, or who made them that way. This all borders on theology, and that is a distinctly painful topic. Better to go have a beer and watch a sporting event on the large-screen TV. But this is *not* the view that a mathematician takes.

The world of the mathematician is a world that we have built for ourselves. And it makes good sense that we have done so, for we need this infrastructure in order to pursue the truths that we care about. But the nature of our subject also sets us apart from others—even from close allies like the physicists. We not only have a divergence of points of view, but also an impasse in communication. We often cannot find the words to enunciate what we are seeing, or what we are thinking.

In fact it has taken more than 2,500 years for the modern mathematical mode of discourse to evolve. Although the history of proof is rather obscure, we know that the efforts of Thales and Protagoras and Hippocrates and Theaetetus and Plato and Pythagoras and Aristotle, culminating in Euclid's magnificent *Elements*, have given us the axiomatic method and the language of proof. In modern times, the works of David Hilbert and Nicolas Bourbaki have helped us to sharpen our focus and nail down a universal language and methodology for mathematics (see [KRA] for a detailed history of these matters and for many relevant references). The idea of mathematical proof is still changing and evolving, but it is definitely part of who we are and what we believe.

The discussion in the next section sheds further light on these issues.

8 Plato vs. Kant

It is by logic we prove, it is by intuition that we invent.

—Henri Poincaré

A debate has been festering in the mathematics profession for a good while now, and it seems to have heated up in the past few years (see, for instance [DAV]). And the debate says quite a lot about who we are and how we endeavor to think of ourselves. It is the question of whether our subject is Platonic or Kantian.

The Platonic view of the world is that mathematical facts have an independent existence—very much like classical Platonic ideals or Forms—and the research mathematician *discovers* those Forms. These forms are like archetypes for the objects that we actually see in the real world (according to the traditional Platonic point of view). Our ideas derive from those. It should be clearly understood that, in the Platonic view, mathematical ideas exist in some higher realm that is independent of the physical world, and certainly independent of any particular person. Also independent of time. The Platonic view poses the notion that a theorem can be "true" before it is proved.

The Kantian view of the world is that the mathematician creates the subject from within himself/herself. The idea of set, the idea of group, the idea of pseudoconvexity, are all products of the human mind. They do not exist out there in nature. We (the mathematical community) have *created* them.

My own view is that both these paradigms are valid, and both play a role in the life of any mathematician. On a typical day, the mathematician goes to his/her office and sits down and thinks. He/she will certainly examine mathematical ideas that already exist, and can be found in some paper penned by some other mathematician. But he/she will also cook things up from whole cloth. Maybe create a new axiom system, or define a new concept, or formulate a new hypothesis. These two activities are by no means mutually exclusive, and they both contribute to the rich broth that is mathematics.

Of course the Kantian position raises interesting epistemological questions. Do we think of mathematics as being created by each individual? If that is so, then there are hundreds if not thousands of distinct individuals creating mathematics from within. How can they communicate and share their ideas? Or perhaps the Kantian position is that mathematics is created by some shared consciousness of the aggregate humanity of mathematicians. And then is it up to each individual to "discover" what the aggregate consciousness has been creating? Which is starting to sound awfully Platonic. Saunders Mac Lane [MAC] argues cogently that mathematical ideas are elicited or abstracted from the world around us. This is perhaps a middle path between the two points of view.

The Platonic view of reality seems to border on theism. For if mathematical truths have an independent existence—floating out there in the ether somewhere—then who created those truths? And by what means? Is it some higher power, with whom we would be well-advised to become better acquainted?

The Platonic view makes us more like physicists. It would not make much sense for a physicist to study his subject by simply making things up. Or cooking them up through pure cogitation. For the physicist is supposed to be describing the world around him/her. A physicist like Stephen Hawking, who is very creative and filled with imagination, is certainly capable of creating ideas like "black hole" and "supergravity" and "wormholes," but these are all intended to help explain how the universe works. They are not like manufacturing set theory or recursive functions.

There are philosophical consequences for the thoughts expressed in the last paragraph. Physicists do not feel honor-bound to prove the claims made in their research papers. They frequently use other modes of discourse, ranging from description to analogy to experiment to calculation. If we mathematicians are Platonists, describing a world that is "already out there," then why cannot we use the same discourse that the physicists use? Why do we need to be so wedded to proofs? This is a query that demands serious consideration. Our mathematical world is definitely "out there," insofar as it is a logical consequence of what we do. But is that Platonic, in the sense that it has existence independent of ourselves, or is it Kantian, in the sense that it emanates from us?

One can hardly imagine a sociologist trying to decide whether his/her discipline is Platonic or Kantian. Nor would a physicist ever waste his/her time on such a quest.[9] People in those disciplines know where the grist of their mill lives, and what they are about. Physicists use mathematical abstraction, but they seem to know how to put it into context. The questions posed here do not really make sense for them. We mathematicians are somewhat alone in this quandary, and it is our job to take possession of it. If we can.

It appears that sociologists and physicists are certainly Platonists. What else could they be?[10] It is unimaginable that they would cook up their subject from within themselves. Certainly philosophers can and do engage in this discussion, and they would also be well-equipped (from a strictly intellectual perspective) to engage in the Platonic vs. Kantian debate. But they have other concerns. This does not seem to be their primary beat.

The article [MAZ] sheds new and profound light on the questions being considered here. This is a discussion that will last a long time, and probably will never come to any clear resolution.

Once again the Platonic vs. Kantian debate illustrates the remove that mathematicians have from the ordinary current of social discourse. How can the layman identify with these questions? How can the layman even care about them? If I were a real estate salesman or a dental technician, what would these questions mean to me?

9 Seeking the Truth

> In what we really understand, we reason but little.
>
> —William Hazlitt

Mathematicians are good at solving problems. But we have recognized for a long time that we have difficulty with communicating with laymen, with the public at large, with the press, and with government agencies. We have made little progress in treating this particular conundrum. What is the difficulty?

Part of the crux is that we are not well-motivated. It is not entirely clear what the rewards would be for vanquishing this puzzle. But it is also not clear what the methodology should be. Standard mathematical argot will not turn the trick. Proceeding from definitions to axioms to theorems will, in this context, fall on deaf ears. We must learn a new *modus operandi*, and we must learn how to implement it.

This is not something that anyone is particularly good at, and we mathematicians have little practice in the matter. We have all concentrated our lives in learning how to communicate *with each other*. And such activity certainly has its own rewards. But it tends to make us blind to broader issues. It tends to make us not listen, and not perceive, and not process the information that we are given. Even when useful information trickles through, we are not sure what to do with it. It does not fit into the usual infrastructure of our ideas. We are not comfortable processing the data.

---

[9] Although string theorists may be pushing the envelope for this question. They have produced over 1000 papers on a theory that does not know in which dimension it lives, and that has produced no verifiable formula or assertion. So string theory, at least today, is arguably Kantian. And that is quite different from most modern physics.

[10] Although a physicist may put a finer point on it and assert that he/she has no care for a Platonic realm of ideas. Rather, he/she wishes to run experiments and "ask questions of nature."

This is our own fault. This is how we have trained ourselves, and it is how we train our students. We are not by nature open and outreaching. We are rather parochial and closed. We are more comfortable sticking close to home. And, to repeat a tired adage, we pay a price for this isolation.

10 Brave New World

For most wearers of white coats, philosophy is to science as pornography is to sex: it is cheaper, easier, and some people seem, bafflingly, to prefer it. Outside of psychology it plays almost no part in the functions of the research machine.

—Steve Jones

For the past 2,500 years, mathematicians have enjoyed a sense of keeping to themselves, and playing their own tune.[11] It has given us the freedom to think our own thoughts and to pursue our own truths. By not being answerable to anyone except ourselves, we have been able to keep our subject pure and insulated from untoward influences.

But the world has changed around us. Because of the rise of computers, because of the infusion of engineering ideas into all aspects of life, because of the changing nature of research funding, we find ourselves not only isolated but actually cut off from many of the things that we need in order to prosper and grow.

So it may be time to re-assess our goals, and our milieu, and indeed our very *lingua franca*, and think about how to fit in more naturally with the ebb and flow of life. Every medical student takes a course on medical ethics. Perhaps every mathematics graduate student should take a course on communication. This would include not only good language skills, but how to use electronic media, how to talk to people with varying (non-mathematical) backgrounds, how to seek the right level for a presentation, how to select a topic, and many of the other details that make for effective verbal and visual skills. Doing so would strengthen us as individuals, and it would strengthen our profession. We would be able to get along more effectively as members of the university, and also as members of society at large. Surely the benefits would outweigh the inconvenience and aggravation, and we would likely learn something from the process. But we must train ourselves (in some instances *re*-train ourselves) to be welcoming to new points of view, to new perspectives, to new value systems. These different value systems need not be perceived as inimical to our own. Rather they are complementary, and we can grow by internalizing them.

Mathematics is one of the oldest avenues of human intellectual endeavor and discourse. It has a long and glorious history, and in many ways it represents the best of what we as a species are capable of doing. We, the mathematics profession, are the vessels in which the subject lives. It is up to us to nurture it and to ensure that it grows and prospers. We can no longer do this in isolation. We must become part of the growing and diversifying process that is human development, and we must learn to communicate with all parts of our culture. It is in our best interest, and it is in everyone else's best interest as well.

---

[11] Although I would be remiss not to note that Archimedes, Newton, and Gauss were public figures, and very much a part of society.

References

[ATI] M. Atiyah, Beauty in Mathematics, video.google.com/videoplay?docid=-5911099858813393554

[DAV] E. B. Davies, Let Platonism die, *Newsletter of the European Mathematical Society* 64(1007), 24–25.

[DYS] F. Dyson, *Disturbing the Universe*, Basic Books, New York, 2001.

[FER] H. Ferguson, Sculpture Gallery, http://www.helasculpt.com/gallery/index.html

[HOF] D. Hoffman, The computer-aided discovery of new embedded minimal surfaces, *Math. Intelligencer* 9(1987), 8–21.

[KRA] S. Krantz, *The Proof is in the Pudding: A Look at the Changing Nature of Mathematical Proof*, Springer Publishing, to appear.

[MAC] S. Mac Lane, Mathematical models: a sketch for the philosophy of mathematics, *American Mathematical Monthly* 88(1981), 462–472.

[MAN] B. Mandelbrot, *The Fractal Geometry of Nature*, Freeman, New York, 1977.

[MAZ] B. Mazur, Mathematical Platonism and its opposites, http://www.math.harvard.edu/~mazur/

Department of Mathematics, Washington University in St. Louis, St. Louis, Missouri 63130
sk@math.wustl.edu

8

What's a Nice Guy Like Me Doing in a Place Like This?

Alan H. Schoenfeld

Unlike most of the authors in this collection, I'm not actively in the business of proving theorems, and I haven't been for quite some time. That's one reason for the title of this piece. In fact, my primary home is in a school of education. That's another reason. How could someone trained as a mathematician wind up in mathematics education—and not only love it there, but feel good about it and hope to induce other mathematicians to make the trip?

The journey wasn't always smooth. Moving into educational territory includes a shift in identity, which can be difficult for those of us who have been raised to think of mathematics as the most exalted intellectual enterprise available to humankind, and to think of education as something less than fully worthy.[1]

My goals for this essay are to indicate two things. First, there is a very important role for mathematicians to play with regard to mathematics education, at any point in their careers. Second, while it can feel truly traumatic to "leave the faith," it is possible to find pathways in educational research that are every bit as challenging and intellectually rewarding as those in mathematics research—with the bonus that one has the opportunity to make a significant difference in the real world.

In the Beginning

I got my Ph.D. from Stanford (in topology and measure theory, not that the details matter at this point) in 1973. My wife was a couple of years behind me, so I took a position as a

---

[1] When I moved to the University of Rochester in the early 1980s, I moved from a straight appointment in mathematics to a joint appointment in mathematics and education. After I gave a mathematics colloquium related to my work on problem solving, one of my mathematics colleagues said, "Well, somebody's got to do this education stuff. I guess I'm glad it's you." Some years later Leon Henkin, tongue only slightly in cheek, asked a group of mathematicians which of the following we thought was the most revolutionary change during our lifetimes: the change in gender roles over the course of the twentieth century, the breakup of the Soviet union, or the fact that mathematicians were now talking about education at the Annual Meetings. Times have changed, but change has been slow.

lecturer at U.C. Davis while she finished up. I had great fun doing mathematics those two years. Admittedly, there was a slight feeling of guilt: being paid for doing something I liked doing so much was sort of like being paid for doing New York Times crossword puzzles. At the same time, I truly enjoyed my teaching: my teaching ratings were close to those of colleagues who had won campus-wide distinguished teaching awards.

Interestingly, that appeared to be a problem. Despite having published a fair number of papers, I was told by more senior colleagues that I risked being viewed as "just a teacher" if I paid too much attention to my teaching. Some advised me to skip my office hours and go home to prove more theorems. That's what would establish my career.

When I mentioned this tension to my friend Ruth von Blum, she said that I should talk to Fred Reif at Berkeley. Reif was a physicist by training, and he had been one of the founders of the Graduate Group in Science and Mathematics Education at Berkeley. SESAME, as it was known, drew senior faculty from mathematics and the sciences who had an interest in education. And, Ruth said, they went about educational work scientifically. I had no idea what that meant, but Fred and Ruth set about trying to convince me that there was a principled (rather than ad hoc) way to go about doing educational research. Fred argued that education at the time (1975) was in much the same state as medicine had been at the turn of the twentieth century—in essence, a folk art. Abraham Flexner's 1910 report "Medical Education in the United States and Canada" had decried the state of contemporary medical practice and argued that it was time to establish the practice of medicine on a scientific footing. Soon afterward the first teaching hospital was founded at Johns Hopkins, and medical practice evolved as the twentieth century passed. Reif argued that the emerging field of cognitive science, which drew upon artificial intelligence, education, linguistics, philosophy and psychology, had the potential to provide a basis for education in the same ways that the study of anatomy and physiology had provided a basis for medical practice. This was a time to get in on the ground floor, to make a difference as the field found itself. Ruth tried to convince me that this could be a serious business. "How much mathematics do you know?" she asked. I said that I was the world's expert on a very small chunk of mathematics. "And how long did it take you to develop that expertise?" A number of years, I said. "Remember," said Ruth, "Human beings are a lot more complex than symbol systems. It'll take you at least that long before you begin to understand mathematical thinking and learning."

I didn't believe it, of course. But there was a lovely problem I'd thought about working on. I'd recently read Pólya's book *How to Solve It*. Pólya identified a series of problem-solving strategies he called "heuristics"—rules of thumb (with no guarantees!) for getting a handle on tough problems and making progress toward solving them. I'd loved the book ("I do the things Pólya says he does. Hot damn, I must be a real mathematician!"), and I wasn't alone. Generally speaking, mathematicians who read Pólya resonated with what he wrote. But, when I spoke to the coaches of various Putnam exam teams, they all said the same thing: "I love Pólya, but my students don't get any better at problem solving by reading his books." I looked at the math ed literature, and the findings were similar: "I tried to teach à la Pólya. My students and I had a great time, but when we finished I couldn't produce any evidence that they were better at problem solving than they were before."

So I was confronted by a nice challenge. If Pólya was right, then we should be able to teach students to be much better problem solvers. But the evidence said something was missing. What was it?

And I had a job offer. Fred offered me a three-year postdoc, which he promptly advised me not to take. (His rationale: after three years of a postdoc in math education, I would no longer be employable as a mathematician and wouldn't be able to claim formal training in education, so neither mathematics departments nor schools of education would want to hire me.)

In short, my choice was either to accept a reasonable but not very exciting position I'd been offered as a mathematician or to risk becoming an academic exile. The first was a secure bet, the second a shot in the dark. What to do?

The pluses and minuses of remaining a mathematician. I love mathematics, and I'm not bad at it; I could've settled happily into a life of theorem-proving. But when I asked the hard question, "Would mathematics miss me if I disappeared?" the honest answer was no. The truth was that I wasn't very likely to re-shape the field. And, if they were really needed, any theorems that I might have proven would have been proven by someone else. Note that this self-evaluation isn't simply a statement about me; it's a statement about the relationship of anyone to the state of the field. Mathematics has a distinguished history of more than 2000 years. What are the odds of any individual making a radically new contribution?

The pluses and minuses of moving into education. For one thing, mathematics education needed (and needs!) honest-to-goodness mathematicians. Who better than a mathematician to have a feel for what it means to "think mathematically?"[2] Second, the state of mathematics education as a field is very different from that of mathematics. To use Kuhn's term, mathematics has been in a state of "normal science" for some time—in our language, the basis vectors of the space have been well established. Mathematics education, however, is only a few decades old (The *Journal for Research in Mathematics Education* was first published in 1970), and its scientific foundations (e.g., the emerging field of cognitive science) are still very much in flux. There was much more room for foundational contributions. And, there's a history of "crossovers" making contributions in the fields they move into. Besides, I'd get to live in Berkeley for three years.

So, I thought I'd give math ed/cognitive science a few years, and see if I could clean up problem solving. If it didn't work (or even if it did, but it wasn't as much fun as I hoped) then I'd think about returning to full-time mathematics. Little did I know that I'd be fully engaged, thirty-five years later, in the study of issues that are every bit as challenging, though not as "clean," as mathematical phenomena.

Just What is Math Ed?

So, just what do mathematics educators do? And can any of it be rigorous, or is it all just BS?

Despite the one-liner about mathematicians being machines that turn coffee into theorems, there's a great deal of diversity in what mathematicians do. Of course we prove

---

[2] This statement doesn't mean that such understandings come automatically, or that others can't have them. Some athletes or opera singers become coaches, some great and some not; and some great coaches weren't great athletes or singers. But, there are significant dangers when people who don't have a sense of what it is to do mathematics—to engage in mathematics as an act of sense-making—are shaping kids' mathematical instruction.

theorems (and more deeply, develop theory); but mathematics runs the gamut from (very) pure to (very) applied. Sometimes, the applications even come as a surprise—it's pretty safe to say that George Boole would be amazed at current computational applications of Boolean Algebra! Boolean Algebra was a piece of pure mathematics, right?

Well, actually not. Although Boole was a mathematician to the core, some of his fundamental work was aimed at axiomatizing... the laws of thought! In 1847 Boole published *The Mathematical Analysis of Logic, being an essay toward the calculus of deductive reasoning*. That book did indeed establish the foundations for Boolean algebra and the propositional calculus. However, that volume was subsumed, seven years later, by the more ambitious (1854) volume *An investigation of the Laws of Thought, on which are founded the mathematical theories of logical and probabilities*. Boole is a noble antecedent for those of us who try to do applied work with mathematical rigor.

Before getting down to details, let me offer a framing that I find useful for thinking about the relationship between pure and applied work. I have no idea how well this is known within the mathematical community.

When I was growing up as a mathematician, the mathematical circles I traveled in reveled in these famous words from G. H. Hardy's (1940) *A Mathematician's Apology*:

> I have never done anything "useful." No discovery of mine has made, or is likely to make, directly or indirectly, for good or ill, the least difference to the amenity of the world. I have helped to train other mathematicians, but mathematicians of the same kind as myself, and their work has been, so far at any rate as I have helped them to it, as useless as my own. Judged by all practical standards, the value of my mathematical life is nil; and outside mathematics it is trivial anyhow. (p. 49)

That is *so* mid-twentieth century Cambridge—perhaps in an attempt to one-up the "other" culture as described by C.P. Snow (1959):

> The climate of thought of young research workers in Cambridge then was not to our credit. We prided ourselves that the science we were doing could not, in any conceivable circumstances, have any practical use. The more firmly one could make that claim, the more superior one felt.

This is certainly a statement about the primacy of pure mathematics over applied mathematics (if not everything else—mathematics is, after all, the Queen of the sciences).

In fact, the primacy of pure over applied was deeply a part of American culture as well. During World War II, Vannevar Bush served under President Franklin Delano Roosevelt as head of the U. S. Office of Scientific Research and Development. Roosevelt asked Bush to make a plan for American scientific R&D following the war. Bush's report, *Science, the Endless Frontier*, was a proposal (essentially a blueprint) for establishing the National Science Foundation.

Bush was clearly a purist, emphasizing the importance of basic research. He claimed that applications followed theory, in that it was not possible to predict in advance where basic research would lead, but that fundamental findings were the source of myriad applications. The sequence Bush envisioned, and which is still common, is given in Figure 1:

Basic Research → Applied Research → Development → Use in Practice

Figure 1. The progression from research to use (after Stokes, 1997, p. 10).

There are, of course, archetypes of scientists whose careers can be seen as exemplifying either the pure or the applied end of the spectrum. Donald Stokes (1997) identifies Niels Bohr and Thomas Edison as archetypal basic and applied researchers, respectively. But Stokes claims that a 1-dimensional spectrum, with examples such as Bohr and Edison at the poles, is too restrictive. He invokes Louis Pasteur, whose framing of many problems was fundamental (e.g., the germ theory of disease) while being motivated by, and having direct applications to, significant applied problems (e.g., food spoilage and curing diseases such as anthrax, cholera, rabies, and tuberculosis). Rather than plunk Pasteur down in the middle of a spectrum with Bohr and Edison at the ends, Stokes argues that it would be better to view things along two dimensions, *Quest for Fundamental Understanding* and *Considerations of Use.* This leads to the two by two matrix in Figure 2:

	Considerations of Use?	
	Yes	No
Quest for Fundamental Understanding? — Yes	Pure Basic Research (e.g., Bohr)	Use-Inspired Basic Research (e.g., Pasteur)
No		Pure Applied Research (e.g., Bohr)

Figure 2. A two-dimensional representation of "basic" and "applied" considerations for research (after Stokes, p. 73).

Pasteur fits comfortably into this scheme—but more importantly, the representation shows how theory and practice can be mutually supportive, if not synergistic. It's that kind of synergy I look for in educational research.

Let me turn to some examples, first to indicate the nature of the enterprise and then to briefly describe some of my own work. The idea here is to challenge some preconceptions readers may have, both about human behavior and about the kinds of work that can be done in the field. So, let me start with a simple axiom:

> Humans don't perceive reality directly. Rather, we interpret the sensory input we receive on the basis of certain "interpretive filters" that we build on the basis of our experience.

The fact that we don't perceive reality directly should come as no surprise—after all, we fall prey to optical illusions. So, there are some intervening psychological mechanisms. But interpretive filters? That is, we learn to "see" things in particular ways. As a case in point, you may have heard about the Amazon forest dwellers who, never having seen any objects at a distance because of the dense forest, had to learn that in open space, animals at a distance weren't smaller; they just seemed that way.[3]

Here's an example closer to home. Figure 3 shows the graphs of two functions, which seem to approach each other as $|x|$ gets large:

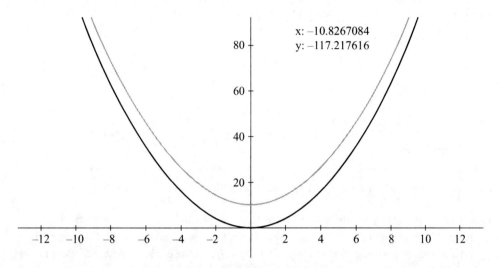

Figure 3. The graphs of two functions.

Note that I say "seem." If I'd started by telling you that I (well, actually, google) had produced the graphs of $y = x^2$ and $y = x^2 + 10$, you'd have imagined the lower graph being "lifted" up by 10 units to produce the upper graph. And once you know that the upper graph is supposed to be a vertical translation of the lower one, you can begin to see (or trace) the constant distance between the two curves. This is a case of your mathematical knowledge ultimately overcoming your perception—humans tend to perceive distance via perpendicular lines, and if you draw the perpendiculars from the top graph to the bottom one, the distances do get smaller as $|x|$ increases. Part of becoming good at mathematics involves learning to pick up various perceptual clues that are unavailable to others who don't have the same experiential base (see, e.g., Stevens and Hall, 1998).[4] Here's a case in point.

Many years ago a student of mine built a "discovery" curriculum for exploring linear functions. She didn't want to simply define slope and show how it worked; she wanted students to develop a "feel" for it. So, she put together some worksheets, which

---

[3] Colin Turnbull (1961) relates the story of an African forest dweller, upon leaving the forest, seeing a herd of buffalo at a distance and asking what kind of insects they are. This story sounds a bit too good to be true, but it's also too good not to pass along.

[4] There are a lot more examples where this comes from. The anthropologist Charles Goodwin (1994) talks about "professional vision," in which members of a discipline learn to see things much the same way as each other, but in ways that are essentially invisible to non-professionals. For example, a slightly mottled and shaded patch of dirt may mean nothing to you or me but, to an archaeologist, is clearly the site of a "post mold"—the decayed remnants of a wooden post that once supported a house.

students would use on some first-generation computers (Apple IIs). Here was one of the worksheets:

> Clear the screen and type in these equations, one at a time.
>
> $$y = 2x + 1$$
> $$y = 3x + 1$$
> $$y = 4x + 1$$
>
> What do you notice?
>
> What stays the same?
>
> What changes?
>
> What do you think will happen if you type in $y = 5x + 1$?
>
> Sketch your prediction on this empty graph [given below in the original] and then try it on the computer.

Figure 4. A "guided discovery" task. With permission, from Schoenfeld, 2009.

Now you could pretty much do this task with your eyes closed. Each of the graphs passes through the point (0,1), and the lines get steeper as the coefficients of x (their slopes) increase. But naive students don't see what we see, and the student responses were not at all what my student expected. Here are the relatively typical responses given by one pair of students:

> What do you notice? *The higher the number you are multiplying by x the more upright the line.*
>
> How are these lines similar? *All practically the same angle.*
>
> How are they different? *They're not the same angle.*
>
> What do you think will happen if you type in $y = 5x + 1$? Sketch your prediction on this empty graph and then try it on the computer. [The student sketch is given in Figure 5.]
>
> What happened? *It went more upright than the other lines but less upright than we thought it would be.*

In short, these students saw very different things than my student expected them to see. The students were facing a graph showing that the lines $y = 2x + 1$, $y = 3x + 1$, and $y = 4x + 1$ all pass through the point (0,1)—yet their sketch passes through the point (0,5). This was not atypical. Students failed to see things we consider to be almost impossible to ignore, and they often fixed on things that we consider inconsequential (e.g., noting that as the coefficient of x increases, the line "starts further to the right" (if you think of the line being sketched from the bottom to the top of the screen). Odd as it may seem, much of our perception (at least, what we attend to in perception) is *learned*.

And, of course, what students perceive can be wrong. We're all familiar with the students who write $(a + b)^2 = a^2 + b^2$. But what matters more is that there can be remarkable consistencies in student misperceptions—which then lead to opportunities for instruction.

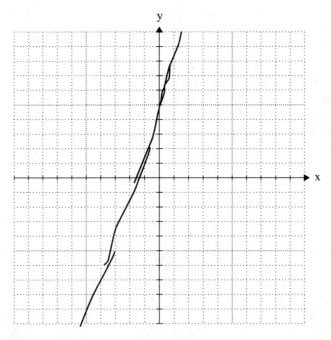

Figure 5. The student sketch. With permission, from Schoenfeld, 2009.

Two of my favorite papers in the literature begin to show how one can make a science of what, until now, has been illustrated by a series of anecdotes. The first is a famous piece by George Miller (1956). In a series of studies, Miller observed that people's ability to perform a range of mental acts were circumscribed in various ways: there were limits to the number of things they could keep "in mind" (technically, in "short term memory" or STM) as they engaged in mental activity.

A simple "proof": Take a second to memorize the two three digit numbers 379 and 658. Now, close your eyes and try to multiply them.

The odds are that you found it impossible, or that if you succeeded, you did so by repeating the results of sub-operations until you had them memorized and could work with them. You're not alone: I've used this task in talks to thousands of mathematicians at various mathematics meetings, and nobody has produced an answer. The reason is that there's too much to keep track of: the full multiplication requires keeping track of a dozen or so subtotals. Miller's paper, entitled "The magic number seven, plus or minus two: some limits on our capacity for processing information," indicated that the vast majority of people are capable of juggling between five and nine pieces of information in their heads, and the multiplication task makes larger demands on cognitive processing. (The way we deal with more information is by "chunking," making units that only occupy one space in STM. For example, we chunk the three digits in our telephone area code into one unit. We also build "recognition vocabularies" that allow us to recognize and respond to collections of things as though they were one unit. Our recognition vocabularies are quite large, and they're not just for words.

For many years, I wondered why I wasn't a good chess player. I'm a pretty good strategist, but when I did play (which was rare) I was often beaten by people whom I knew weren't nearly as good strategically as I was. The reason, it turns out, is that people who play a lot of chess develop a visual vocabulary of tens of thousands of chess board positions—and react to them as automatically as you put your foot on the brake when you see a red octagonal sign while you're driving (it's part of your recognition vocabulary). This phenomenon explains why experienced chess players can play forty simultaneous games—they're not thinking each move through, but simply "reading" the positions and responding accordingly as they move from board to board. (And, it explains why students may be stymied by expressions such as $e^{i\pi} = -1$, while we don't even blink.) Indeed, the going estimate for the development of expertise in any field is some 5000–10,000 hours of concerted practice.

The second of my favorite papers has the exciting title "Diagnostic models for procedural bugs in basic mathematical skills" (Brown and Burton, 1978). Brown and Burton spent a substantial amount of time looking at children's arithmetic work. They found substantial consistencies in the children's (mis-)understandings. For example, see what you can say about this boy's work on six subtraction problems:

$$\begin{array}{cccccc} 278 & 352 & 406 & 543 & 510 & 1023 \\ -135 & -146 & -219 & -367 & -238 & -835 \\ \hline 143 & 206 & 107 & 176 & 272 & 88 \end{array}$$

The first observation is that he got four out of six right, for a score of 67%. A second observation is that he has trouble with zeros—they're involved in the fourth and sixth problems, which he got incorrect. But, he got the fifth problem correct, so he doesn't "just" have a problem with zeros. It's more complex than that.

I won't explain the child's difficulty, but I'll offer you a challenge. Given what you have seen thus far, can you predict the answer the student will give to the problem

$$\begin{array}{r} 605 \\ -237 \end{array} \ ?$$

Brown and Burton could. In fact, they were able to develop a 16-item diagnostic test that allowed them to predict, about half the time, the *incorrect* answers that students would get to subtraction problems—before the students worked the problems! (Given that a non-trivial percentage of errors are slips, the 50% figure is pretty amazing.)

This is a major result, for both theoretical and practical reasons. For one thing, it means that children's arithmetic errors aren't random; they're systematic. That blows a standard piece of pedagogical wisdom right out of the water. What makes a good teacher? Many people would say it's the person who has half a dozen explanations of any particular idea in her back pocket. If the first one doesn't take, maybe the second will. If the second doesn't take, maybe the third one will, and so on. That perspective depends on the "blank slate" or "empty jug" image of the student, the tabula rasa waiting to be written on or the container waiting for knowledge to be poured in. But, as Brown and Burton's work shows, it's not that the slate is blank or the jug is empty. If you can predict a child's incorrect answer in advance, then that child *has* learned something—and writing over it may not be possible, until what's wrong has been fixed. (This shouldn't be shocking. We're accustomed to this possibility in the case of physical actions. For example, the aspiring violinist who's picked up the wrong fingering needs to "unlearn" the incorrect physical habit, and doing so can be much harder than picking up the right one in the first place. The same is true with regard to cognition.)

The theoretical issue is that the predictive nature of their work demonstrated (back in 1978) that the idea of *cognitive modeling* is a real possibility. If researchers can understand someone's thinking well enough that they can predict, in advance, the incorrect answer that that person will get to a problem—by specifying the incorrect algorithm that the person has developed for him/herself—then they must have a pretty good handle on that person's thought processes. In short, though human behavior is hardly predictable in general, one can set about being scientific in the characterization of human knowledge and behavior. That's what a part of math ed, and a core part of my own work, are all about.

So What Do I Do?

Here comes 35 years of work in a few pages. If this capsule description captures your interest, you might be interested in looking at the book that sums up my early work on problem solving (Schoenfeld, 1985) and the more recent summary of my work on human decision making (Schoenfeld, 2010).

The biographical part of this chapter left off at the point when I decided, in 1975, to take a close look at problem solving. The challenge was, why didn't Pólya's description of problem solving strategies work as a basis for instruction?

My work over the next decade consisted of a combination of observations (just how do experts solve problems?), laboratory studies (if the observational work suggests that effective problem solvers do things in particular ways, can I train students to do the same?), and teaching (what happens as I take the ideas from the research and try to make them work in a real-live instructional context?). Over those ten years, the research and development lived in happy synergy.

First, the close observation of proficient problem solvers revealed why Pólya's ideas hadn't translated well into problem solving instruction. Yes, it's true that when good problem solvers get stuck, they turn to sensible alternatives—e.g., trying easier related problems and then exploiting either the methods or the results of the related problems in working on the original problem. And it's true that mathematicians who read *How to Solve It* and/or Pólya's other books recognized the strategies and said, "yes, that's what I do." But it turned out that, while it was easy to recognize Pólya's descriptions of problem solving strategies once you'd figured out the strategies for yourself, there wasn't enough detail in Pólya's descriptions to allow beginners to use them. For example, the level of description "try an easier related problem" is too coarse. There are at least a dozen different ways to craft easier related problems: considering a smaller n, looking at simpler functions or simpler geometric figures, picking special values or special cases, and so on. Each of these is a strategy on its own merits. Which, if any of these, will unlock this particular problem? One has to choose a direction. Then, once one has chosen the direction, one has to be able to pursue it—to solve the related problem—and to figure out how to exploit either the method (it worked here. How can I use it on the original problem?) or the result (I've gotten part-way there. How do I get the rest of the way?) to solve the original problem. The bottom line: once you decompose the strategy into the dozen sub-strategies and teach students to how to recognize when they're appropriate and how to use them, then students can use them. This was part of my early work.

Second, it turns out that students fail to solve a lot of problems that they should be able to solve because they wind up chasing wild mathematical geese. Typically, students will read a problem, pick a direction to pursue, and then pursue it come hell or high water. If they choose well, they may succeed. If they choose poorly and don't reconsider, then their solution attempt is doomed: you can't make progress in the right direction(s) if you're busy working in the wrong direction(s).

This is far from a minor issue. It turns out that when you ask students to solve unfamiliar problems[5] out of context, more than half the time the graph of their solution looks like Figure 6.

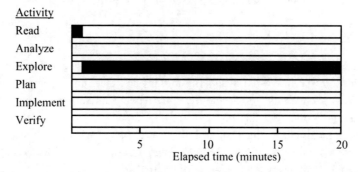

Figure 6. Time-line graph of a typical student attempt to solve a non-standard problem. (From Schoenfeld, 1985, with permission).

Third, and this may sound strange if you haven't heard of it before: students' beliefs about the nature of the mathematical enterprise are powerful shapers of how they go about doing mathematics. I tripped over this somewhat by accident, when I asked college students to solve the problem in Figure 7.

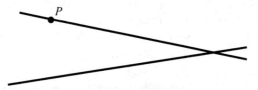

You are given two intersecting straight lines and a point P marked on one of them. Show how to construct, using straightedge and compass, a circle that is tangent to both lines and has the point P as the point of tangency to the top line.

Figure 7. A geometry problem. (From Schoenfeld, 1985, with permission).

Typically, students would intuit that the other point of tangency would be the same distance from the point of intersection of the two lines as it is from P—and then they would

---
[5] Obviously, if you're testing the content of a particular chapter from their textbook, they'll expect to use the knowledge from that chapter. What I'm talking about is what happens at some point when you give them a problem out of the blue—one for which they know the relevant mathematics, but don't have contextual cues about which mathematics to apply to the problem.

conjecture that the line joining *P* and its mirror image would be the diameter of the desired circle. There was a compass on the desk; they'd pick it up and try their construction. When their attempt didn't look right, they'd typically say something like "The midpoint of the diameter isn't far enough to the left. What if we use the arc from the point of intersection instead of the diameter?" Some would remember that the radius of the desired circle would be perpendicular to the top line at *P*—and then would conjecture that the perpendicular at *P*, extended to the bottom line, would be the diameter of the desired circle.

I was shocked. It appeared that the students knew no geometry! I was even more shocked the following year. That year I started with proof problems, e.g., asking students to show that the center of a circle tangent to two given lines lies on the bisector of the angle formed by the two lines. They did so by noting that the center of the circle lay on the perpendiculars from the two points of tangency, and showing that the two triangles formed by the point of intersection, the center of the circle, and the two points of tangency were congruent. I then gave them the construction problem in Figure 7—and they conjectured that the line segment between the point *P* and its mirror image would be the diameter of the desired circle!

To cut a long story (and a couple of years of research) short, it turned out that the origins of the students' strange behavior was in the way they'd been taught high school geometry. They'd done proofs, but the proofs were always of things that were intuitively obvious; they'd never actually used proof as a tool for discovery. And when they did constructions, the standard they were held to was purely empirical: a construction was graded as being correct if it contained the right collection of marks (line segments and arcs) and came quite close to being what it was supposed to be. (That is, an angle bisector had better look like it bisects the angle, an inscribed circle in a triangle had better be tangent to the three sides of the triangle, and so on.) If things looked good, with little tolerance for error, students got full credit. What the students abstracted from their experience (in many different schools—I saw the evidence in high school instruction when I visited classes, but my students came from all over) were two core (though unarticulated) beliefs: (1) proof has nothing to do with mathematical discovery, and (2) the standard of correctness for construction problems is purely empirical. Thus, given a construction problem for which they had to "discover" the answer, they ignored their proof-related knowledge and approached it (and judged their attempts!) by purely empirical standards.

There are lots of other stories—of elementary school students saying that you can get five two-foot boards for a book shelf by cutting up two five-foot boards, or that the number of buses needed to take soldiers to an army base is "31 remainder 12." Such nonsense answers come from another belief that the students developed—that the word problems they study in math class are "cover stories" for simple arithmetic, so their job is to read the problems, abstract the math, perform the operations, and write down the answers. (See Schoenfeld, 2012, for slightly more detail.)

In short, not only was students' mathematical knowledge important in problem solving, but in addition their ability to use problem solving strategies, their decision-making regarding the use of the resources at their disposal (including time!), and their beliefs about the nature of mathematics all shaped their problem solving success or failure. Over the years, I taught various versions of my problem solving course. These incorporated attempts to work through the ideas, sometimes successfully, sometimes not; but the failures taught me as much as the successes, and often sent me back to the lab with insights about what had and hadn't worked. I note that this line of research allowed me to continue a form of mathematical involvement—I was thinking about mathematical thinking all the time—while I was doing work that was intellectually challenging and had the potential to improve mathematics teaching.

Although my research did explain the sources of success and failure in problem solving, it didn't yet explain why people made the choices they made while engaged in problem solving. Moreover, as an act of decision making, problem solving is theoretically simple. There's one individual, alone, working on a task that doesn't change while he or she is working on it. Contrast that with teaching. It too is an act of problem solving, in the broadest sense. (You can think of problem solving as a goal-oriented activity, the goal being to solve the problem you confront. Teaching, like most human activity, is also goal-oriented. Among the goals a teacher is trying to achieve at any given time is the goal of having students develop an understanding of particular content.) But teaching is inherently social and highly contingent—for example, you may think things are going along swimmingly when a student says something and you realize that he or she (and quite possibly half the class) don't have the understandings you'd assumed they have. Oops, time for plan B!

The challenge I've worked on for the past twenty-five years has been to model the act of teaching. Here I mean modeling in a sense akin to that of applied mathematics. Whether one is building a model of something as simple as the steady state conduction of heat in a flat plate, of the infrastructure and control systems of a nuclear power plant, or of a teacher working with a group of high school students, the basic questions one has to address are the same. One needs a theory of what counts, how those things are related (that is, how the various components of the system affect each other), and how they change in time. Then, to model any particular situation (say, for example, when a source of heat at constant temperature is applied to a particular point on the flat plate mentioned above; or there is a leak in a part of the plant's cooling system; or a student says something unusual) one enters those inputs into the model, runs it, and sees what happens. If the model is good, then the behavior of the model should correspond pretty well to the behavior of the system being modeled. And if the theory produces a range of models that correspond to a wide range of situations, then there is a fair chance that the theory captures a significant part of what counts.

To give a trivial physical example, consider a theory of gravitational attraction. You'd expect it to be general in the sense that a theory that only allows you to model two-body gravitational attraction wouldn't be very exciting. A decent theory would allow you to characterize the attractions of a fairly broad class of objects, two-body systems among them. The test of the theory is its accuracy over a range of circumstances. So, you'd apply it to our solar system. The theory says that certain things matter: the mass, position, speed, and direction of the objects in the system, and the gravitational attractions between them. You assign the values for the sun and each of the planets at a particular time (it happens to be New Year's day 2012; that's a reasonable start time) and then run the model. If the model predicts that the objects will be in various places at various times, and the predictions are accurate, then you have some faith in the model. If the theory allows you to model a wide range of planetary systems accurately, then you have increasing confidence in the theory.

It's the same in the case of modeling teaching. The first question is, what counts? In simplest terms, the answer is: the teacher's knowledge (of mathematics, of teaching strategies, of the students in the class, of what's been covered, etc.), beliefs (about mathematics, about the students, etc.),[6] and goals (for the lesson, for the students, etc.). Then there has to be a decision-making mechanism. In this case there are two, for typical and atypical situations respectively. A typical situation is one for which a teacher has a routine he or she can call on to deal with the situation—e.g., for collecting homework papers, having students go to the board, for laying out content, and even for dealing with specific student misconceptions, such as when a student writes $(a + b)^2 = a^2 + b^2$. If the teacher has well developed routines, the model of the teacher simply implements those routines when the applicability conditions for them occur.

Here's an example of a non-routine situation (see chapter 5 of *How We Think* for the full detail). Jim Minstrell, winner of a U. S. Presidential teaching award, is teaching at the beginning of the term. His overarching goal for the opening series of lessons is to give students a sense that formulas and their application are not arbitrary, but that their use should make sense. He and his students have talked about various ways to "capture" a particular thing with a number—for example, Olympic sports scores (where the high and low scores are not counted), in descriptions of a population, and so on. He has had eight students measure the width of a table in the classroom. The values they came up with were

$$106.8, 107.0, 107.0, 107.5, 107.0, 107.0, 106.5, 106.0 \text{ cm}.$$

Minstrell's question was, what is the best number to represent the width of the table? Things proceeded according to routine (and can be easily modeled as such) as one student suggested that the class average the eight numbers and another suggested the mode. But then one student said,

> This is a little complicated but I mean it might work. If you see that 107 shows up four times, you give it a coefficient of 4, and then 107.5 only shows up one time, you give it a coefficient of 1, you add all those up and then you divide by the number of coefficients you have.

This is clearly out of the ordinary. And, there is a wide range of possible responses. I know teachers who might respond in each of the following ways:

"That's a very interesting question. I'll talk to you about it after class."

"Excellent question. I need to get through today's plans so you can do tonight's assigned homework, but I'll discuss it tomorrow."

"That's neat. What you've just described is essentially the 'weighted average.' Let me briefly explain how you can write a formula using coefficients similar to the way you've described, and that gives the same numerical result as the average."

"Let me write that up as a formula and see what folks think of it."

"Let's make sure we all understand what you've suggested, and then explore it."

---

[6] Recall the impact of students' beliefs about mathematics on their mathematical performance. The same is true of beliefs related to teaching and their impact.

Each of these answers has pluses and minuses, the magnitude of which depend on what the teacher considers important—different teachers will place different values on things such as keeping the lesson on track, staying on safe ground, exploring new territory, and responding to student initiative. In essence, the way you can model a teacher's decision making in circumstances such as this is to identify the various options (such as those listed above) and, for each, compute the subjective expected utility of the option depending on how important the teacher considers each of the outcomes that would result from taking that option. For example, a teacher who is afraid to go out on a limb and places high value on getting through the day's lesson plan would score high on the first option and low on the last. In contrast, Minstrell scores highest on the last option by far—so that is the option the model says is most likely to take.

As I've noted, it took about twenty-five years to work out the details of this study, and to build enough models of different people, in different circumstances, to be confident that the theory is robust. Like the work on problem solving, it has potential applications. Simply put, the more we know about how something works, the more we're in a position to improve it. And, one of my long-term goals is to find ways to improve mathematics teaching.

Why and How Should You Consider Being Involved in Mathematics Education?

For me, the move from mathematics to education has been tremendously rewarding. Although I am no longer proving theorems in mathematics, I still very much move in a mathematical universe. Thinking about thinking, especially mathematical thinking, is fun; figuring out how to do such work with something approaching scientific rigor has been an interesting and rewarding challenge. To the degree that my work is successful, I help to create a context in which an increasing number of children can experience the pleasures of mathematical thinking and problem solving.

I want to close by noting that there is a role for mathematicians, at all stages of their careers, in mathematics education. There are various ways to dip one's toes into educational waters, often by volunteering at first—whether in local schools, by reviewing (journals and organizations like the NSF always welcome thoughtful new reviewers), or by making oneself available as a partner in curriculum development, co-teaching courses in mathematics education, or being the "outside" member of students' oral exams.

One major reason for this is that there *is* a mathematical way of thinking, a way of mathematical sense-making. Sometimes this kind of sense-making is embodied in instruction, but often it's not. The examples I gave above of students giving physically impossible answers to simple word problems are cases in point. (Also, if you don't know it, you might be interested in reading Max Wertheimer's classic (1959) description of a geometry class in which the students could derive the standard formula for the area of a parallelogram and use it on a series of exercises, but were then completely stymied when he drew a parallelogram in non-standard position (Figure 8).)

Mathematics is wonderful in part because it coheres so nicely. Ultimately, doing mathematics is an act of sense making: if you look for the reasons that things fit together you can find them, and then everything just makes sense. It's this sense of mathematics that we're privileged to live as mathematicians, and it's one that few people have access to. Helping

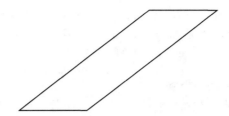

Figure 8. A parallelogram in non-standard position.

others (including our students, and potential teachers) see mathematics this way can help more people to get a glimpse of what it is to engage with mathematics.

At the same time, I urge you to be humble as you reach out. We mathematicians have many things to offer, but so do those we seek to partner with. One of the great things about being a professor is that I get to choose the 80 hours a week that I work—so, when my daughter was in school, I volunteered to be a "mathematical assistant" in her classes. The thing about young children is that they're often honest: when you're working an activity table in a first grade classroom and the students all wander away, your failure to capture them is all too apparent! (I finally got the hang of it and told my cooperating teacher proudly that I'd figured out how to make a particular activity work. She asked if I was ready to take over the whole class. No way! Living in a classroom is a great way to learn just how hard the job of teaching really is.) And it's been many years, but I still remember the very bright second grader who said (as we were working through base 10, toward the end of the year) "You know, Alan, we haven't understood a thing you've said for the past ten minutes."

I don't want to underestimate for a moment the challenges of moving into a new area, especially when it means collaborating with many people who have difficulties with mathematical ideas that may seem trivial to us. There are things they have to learn; there are things we have to learn. And sometimes it's just plain hard. But when it works—and if we keep at it, it will—it's tremendously rewarding.

References

Boole, G. (1847). *The mathematical analysis of logic, being an essay toward the calculus of deductive reasoning*. New York: Philosophical Library.

Boole, G. (1854). *An investigation of the Laws of Thought, on which are founded the mathematical theories of logical and probabilities*. Cambridge: Cambridge University Press.

Brown, J. S. and Burton, R. R. (1978). Diagnostic models for procedural bugs in basic mathematical skills. *Cognitive Science*, 2, 155–192.

Bush, V. (1945). *Science, the Endless Frontier*. Washington, DC: United States Government Printing Office. (Downloaded July 9, 1998, from http://www.nsf.gov/od/lpa/nsf50/vbush1945.htm.)

Flexner, A. (1910). *Medical Education in the United States and Canada: A Report to the Carnegie Foundation for the Advancement of Teaching*. Bulletin Number Four. New York: Carnegie Foundation for the Advancement of Teaching.

Goodwin, C. (1994). Professional Vision. *American Anthroplogist* 96(3), 606–633.

Hardy, G. H. (1940). *A Mathematician's Apology*. Cambridge, England: Cambridge University Press.

Miller, G. (1956). The magic number seven, plus or minus two: some limits on our capacity for processing information. *Psychological Review*, **63**, 81–97.

Schoenfeld, A. H. (1985). *Mathematical Problem Solving.* Orlando, FL: Academic Press.

Schoenfeld, A. H. (2009). Bridging the cultures of educational research and design. *Educational Designer 1(2):* http://www.educationaldesigner.org/ed/volume1/issue2/.

Schoenfeld, A. H. (2010). *How we think: A theory of goal-oriented decision making and its educational applications.* New York: Routledge.

Schoenfeld, A. H. (2012). A modest proposal. *Notices of the American Mathematical Society*, 59(2), 317–319.

Snow, C. P. (1959). *The two cultures.* Cambridge, England: Cambridge University Press.

Stevens, R. and Hall, R. (1998). Disciplined perception: Learning to see in technoscience. In M. Lampert & M. L. Blunk (Eds.), *Talking Mathematics in School: Studies of Teaching and Learning* (pp. 107–149). Cambridge, England: Cambridge University Press.

Stokes, D. E. (1997). *Pasteur's quadrant: Basic science and technical innovation.* Washington, DC: Brookings.

Turnbull, C. (1961). *The forest people.* New York: Touchstone.

Wertheimer, M. (1959). *Productive Thinking.* New York: Harper & Row.

9

A Mathematician's Eye View

Ian Stewart

I have been working in university mathematics for forty years, mainly in the United Kingdom, but with two years in the United States, a year in Germany, and six months in New Zealand.

My current position is slightly unusual: twelve years ago I stopped giving lectures to undergraduates, so that I could focus on "public understanding of science." I did this with the encouragement of my home institution, the University of Warwick, which has always had a remarkably enlightened view of such activities. The research component of my job remained unchanged—except that I wrote more research papers per year after I had stopped lecturing.

My perspective on mathematics at the university level is therefore different, in some ways, from that of a typical mathematician. The work on the public understanding of science involves such things as radio broadcasts, television, writing for newspapers and magazines—and more recently making podcasts and contributing to websites.

What follows is a series of thoughts about particular issues concerning mathematics: both within the subject, and in relation to the outside world.

What Are Your Thoughts on Mathematical Genius?

Some people seem to have a natural ability to think mathematically, at a much higher level of difficulty than others. I do not think that the label "genius" is a good one, however, because it carries a great deal of historical baggage, much of it wrong or simple-minded. It is the experience of most professional mathematicians that some students can understand new ideas very rapidly and develop them in their own minds without much apparent effort, while others struggle with simple concepts and seem unable to grasp what the subject is really about.

These statements contradict the conventional wisdom in psychological circles, which seems to be that there is no such thing as natural talent. Instead, every child is a blank slate, upon which anything can be written, and the only difference between a genius and an ignoramus is a lot of hard work.

As proof of this in the context of mathematics, a study made by Binet around 1900 is often cited. He found that French cashiers in shops were better at arithmetic than most calculating prodigies, but the cashiers came from all walks of life. The usual deduction is that anyone can be trained to calculate fast and accurately, so that "mathematical talent" is solely a matter of hard work and practice.

I think that most professional mathematicians find this argument unconvincing. Certainly I do. We have all encountered students who seem able to live and breathe mathematics with far less effort than others who are struggling despite working enormously hard. Many of us were like that ourselves. I outperformed all of my classmates at mathematics, and did far less work than nearly all of them. I just found it easy. Of course hard work and practice help to develop abilities, and performing at the highest levels requires a lot of sustained effort and experience.

So what—if anything—is wrong with the psychological studies? I think quite a lot is wrong.

The Binet study (at least as it is typically interpreted) involves a clear error in experimental design: claiming that a population is a random sample *after* its members have been chosen to exhibit some particular feature. It is like choosing the tallest people in a room, observing that they come from very diverse backgrounds, and then claiming that anyone can become tall if they put their mind to it. The cashiers were self-selected for their willingness to do lots of arithmetic, and only kept their jobs if they were good at it. The study did show that this ability does not correlate strongly with social background. But that's quite different from "anyone can be trained to do this."

There is also a huge difference between being able to do school-level calculations quickly and accurately, and being a creative mathematician. Most of us can be trained to play, say, a flute, fairly competently. That does not mean that we can be trained to be the new Beethoven. Anyone who runs regularly gets better at it, but that does not imply that we can all become Olympic athletes.

There is a general problem here: a tendency to observe abilities at a moderate level, where improvement through practice is common and expected, and to extrapolate those observations to higher levels of ability. It would be wonderful if we were all born with exactly equal potential at everything, and just had to decide how hard we wanted to work and in what field. But as Steven Pinker argued in his book *The Blank Slate*, the belief that this is the case owes a lot to political correctness and precious little to the evidence.

What Are the Sources of Mathematical Subject Matter?

The main two are the internal demands of mathematics itself, and the demands imposed on it by the outside world. These two ways of motivating mathematics are usually labeled "pure" and "applied" mathematics. The words are not very well chosen: the purity of "pure" mathematics is one of method, not that of the highborn princess who refuses to sully her hands with the impurity of hard work; and not all "applied" mathematics is applied to anything useful (though much is).

An example of internal demands might be the Greek emphasis on logical proof and axioms in geometry, which had little to do with practical applications such as land-measurement. Or, in modern times, the development of topology, which emerged from many areas of mathematics because of a common emphasis on continuity properties. External demands led to such topics as differential equations and numerical analysis.

As these examples—and countless others like them—make clear, these two sources of motivation are not the opposite sides of a coin, but the ends of a continuous spectrum. The creative ideas in mathematics repeatedly wander from internal demands (does a solution to this ODE exist?) to external ones (how can we reconstruct the shape of something from its vibrational spectrum?) and back again. The subject's strength lies in how it combines the two. And almost everything in mathematics owes a lot to sources of inspiration, problems, and ideas, with a very tangled history.

What Are the Human Traits Behind Mathematics?

Among the positive ones are a desire to find answers, a finely honed degree of skepticism about claims lacking adequate evidence, a clear head for logical argument, an ability to recognize a good idea when you see it, an ability to focus on the essentials of a question and to ignore irrelevant or extraneous information, and a total absence of fear regarding anything technical.

Negative ones include a tendency to see the world in black and white and ignore shades of gray; and accepting experimental data, conclusions, and attitudes from the outside world without finding out where they came from and asking whether they are true or defensible.

What Do Mathematicians Mean by "Beauty"?

Not what artists mean. Visual attractiveness (say of some nice fractal) is not what mathematicians mean by beauty, though it is often a sign of it. It is not what artists mean by beauty either, insofar as that concept applies in art nowadays. Both seek something deeper. What mathematicians look for is the structure of the mathematical story in which some particular piece of mathematics is embedded. The result or its proof should be in some sense surprising, but should leave the reader with a sudden feeling of enlightenment, the proverbial little light bulb in the head switching on.

A common source of this kind of beauty is the discovery that a problem arising in one area of mathematics is linked to a totally different area, in a way that no one had previously recognized. The proof of Fermat's Last Theorem, recasting the problem in terms of elliptic curves, is a recent example. Of course such a link alone is not enough—in particular, it must solve the original problem, otherwise it might be just a sterile reformulation.

A proof is a story, told in logic, and the story has to possess a strong sense of plot. If it rambles all over the landscape, or becomes very repetitious, it is not beautiful.

So the essence here is a kind of intellectual beauty, one of ideas and how they fit together. It may appeal to the same overall sense of the esthetic that arises with regard to art, but it differs significantly from the beauty of a painting or a symphony. However, it is not "austere," as Bertrand Russell famously said; just an acquired taste.

Stereotyping and Person Perception, How Are Mathematicians Perceived?

Despite a lot of effort to humanize the image of mathematicians, my experience is that most people think that (a) mathematicians must be very clever, (b) they have not got a clue about the "real world," and (c) it is going to be really difficult talking to them. A recent survey showed that children thought of a typical mathematician as male, elderly, and balding—or maybe with an Albert Einstein haircut.

Most of these stereotypes make little sense if you think about the assumptions behind them. Where do all the old mathematicians come from—spontaneous generation? Stereotypes are hard to erase. Many Americans still think that London suffers from terrible smog (no, the Clean Air Act got rid of the smog fifty years ago).

On the positive side, I did once pass through Immigration in Houston, and the female immigration official looked at my passport, which listed my profession as "mathematician," and said "I love math! It was my best subject at school." And I have found that at parties people no longer say "I was never any good at math at school;" instead they ask "What do you think about chaos theory/fractals/Fermat's Last Theorem?" So maybe the image of mathematics and mathematicians is improving slowly.

If we can break the automatic link with school, and disabuse people of the silly notion that what they did at school was *all the math there is*, it would be wonderful. I have nothing against school math, I hasten to add. But it is not the only place where people's lives are influenced by math. They just think it is.

Mathematics is not the only profession to suffer from such misconceptions. I suspect that most people misunderstand any area of human activity other than their own. I know that many people think that schoolteachers take long summer vacations (no, the students do that) and bank employees go home the moment the doors are closed (no, that is the customers). An entirely intelligent person once remarked to me that "there are an awful lot of cars in the university car park, considering how impoverished the students are supposed to be." I gently pointed out that he was looking at the faculty/staff car parks, and that a University with 15,000 students has around 4,000 staff.

Should Students with High Levels of Ability Get Special Treatment, Just Like Those with Learning Difficulties Do?

To a limited extent and with a lot of thought about safeguards, yes. They should get *some* special treatment. Some people feel this is "elitist." I think this criticism confuses equality of opportunity with equality of achievement.

The 8-year old son of a colleague of mine, while being taught mathematics at a local school, was working his way though several dozen questions of the kind "tell a number story about the number 1." Meaning write down things like "$2 + 1 = 3$" or "$7 - 1 = 6$." After an hour or so he scrawled "$10^{100} - 1 = 999\ldots999$" with the correct number of 9s. No education system should bore students to tears, and destroying the interest of able students is, in my view, close to criminal.

What Are the Interpersonal Consequences and Social Consequences of (Having, Being a) Mathematical Talent?

I have actually had very few problems in those areas. I was a bit shy with girls, as a teenager, but I would not blame my mathematical abilities for it. The main cause was probably that I attended a single-sex school. I am happily married (one marriage, 41 years and going strong) with two sons and three grandchildren. I lived in a village for a time, painted the scenery for the annual pantomime, played on the cricket team, hung around in the pub. Avril and I went to parties and hosted them. I played in a rock band when I was an undergraduate and still own two electric guitars and one acoustic one. Avril and I travel to interesting places (Egypt, New Zealand, Easter Island, the Galapagos, Peru . . .) and share interests in amateur Egyptology and geology; we enjoy the company of like-minded people. I write books that get onto the bestseller lists (sometimes), work with science fiction authors, take part in radio discussions . . . and once introduced a live tiger into a televised lecture.

I have colleagues who own a farm, go hang-gliding, write poetry, fly aircraft, sing in a choir, and own their own companies. There are trained mathematicians in virtually all walks of life.

So I think that relations outside the mathematical community need not suffer as a result of being a mathematician. It depends on what you are willing to get involved with, and how social an animal you are. I admit that I am not especially gregarious, but I don't avoid contact with other people.

Some mathematicians are, I accept, eccentric. Some have little social sense. Some have been clinically insane. But *you do not have to be like that* and most of us are not.

Within the mathematical community, big pluses are its international dimension, and how friendly and unpretentious mathematicians usually are. Mathematicians are mostly normal people who enjoy normal pleasures: we like to go on picnics, visit good restaurants, go to the movies or a sporting event . . . just like real people. Mathematicians can be a bit introspective, especially when working on a research problem. But they do not all suffer from Asperger's syndrome.

What Are Your Thoughts on the Influence of Psychology on Mathematics, Mathematical Discovery, Mathematical Research, or Mathematical Schools of Thought—and Vice Versa?

There do exist different schools of thought in mathematics—such as the pure/applied one alluded to in a previous answer, or the long-running constructivist/nonconstructivist debate. I think that most mathematicians are not greatly bothered by such things—we may have our opinions, and we may even make them public, but we generally feel part of the mathematical mainstream, and are willing to live and let live. I think diversity can be beneficial, and I hate the assumption that there is only one good way to do mathematics (invariably the way the person making that claim does it themselves).

An individual's approach to mathematics is very much a consequence of their personal trajectory, family background, and psychological attitudes. But I do not think there are hard and fast rules for how the link goes. My father was a very practical man, but I much preferred pure math at Cambridge, for instance—before I learned to disregard such classifications. I was the first person in my family to go to University, but I loved it, and thrived. So in those cases, what I did was different from what my background might lead you to expect. On the other hand, I always had wide interests as a child, mainly about something scientific—astronomy, fossils, butterflies—and my interests remain quite broad today. And they are still mostly scientific.

Inspiring researchers often create "schools" of former students, who share their attitudes, work on related problems, and so on. Sometimes this may not be such a good thing—talented students can get stuck in mathematical dead ends, unduly influenced by an enthusiastic but over-specialized thesis advisor who hasn't noticed that the mathematical world has moved on.

What Are Your Thoughts on the Influence of Governments, Politics, History, Economics, Language, Technology, or Religion on Mathematics, Mathematical Discovery, Mathematical Research, or Mathematical Schools of Thought—and Vice Versa?

I am not keen on religion myself, but I doubt that religious beliefs make a big difference to how people do math—except perhaps through its influence on the culture in which they grow up. Some religions tend to stifle original or critical thinking and lead to an over-reliance on authority. I know mathematicians who are Christians of various shades, Orthodox Jews, Jewish atheists, Muslims.... Most are either mild atheists or so moderate that you seldom discover that they are religious. The same goes across the whole of science: it tends to go with having a rational mind.

In Europe, mathematicians tend to be left-wing politically, but there are plenty of counterexamples. Children from wealthy backgrounds tend to go into the business world, not mathematics.

Technology has always had a more influence on mathematical research than we like to think. Many of the great mathematicians of the past invented calculating machines, for instance. Today, when computers have finally become good enough to do serious mathematics, they are having a huge (and on the whole beneficial) influence.

At any given time, some nations seem to excel in mathematics while others falter, and this may to some extent arise from political systems, past history, and so on.

The growing role of Western governments in mathematical research, through organizations like NSF, EPSRC in the UK, and so on, causes me some concern. On the one hand, governments are right to try to guide research into areas where public money is not going to be wasted. On the other hand, too much interference stifles new ideas.

There is some current discussion about US Senators using the legislative process to reject already-funded peer-reviewed research grants in science. This is unacceptable (and usually partisan and small-minded).

Since 1986 the UK government has carried out a regular review of university research (the RAE, Research Assessment Exercise) and this was soon followed by the TQA (Teaching Quality Assessment).

The aim of the RAE was to provide a basis for allocating research funding; to this end it required detailed information about the work of every member of every department. The end result was reasonable, but the same results could have been obtained more easily, and cheaply, by concentrating on a few key indicators. Ironically, the most effective of these (that is, the one that correlated most strongly with the final rating) was the number of research papers published, but this was ruled out on the grounds that using it might lead to too many papers being submitted to journals.

The TQA involved amassing a huge quality of paperwork about procedural issues—specification of a new course, approval of such a course, examination procedures, monitoring student progress, monitoring the procedures for monitoring student progress.... Departments had to provide so much evidence that it typically filled an entire room. Then it was ignored. And important issues such as course content were not examined at all.

These bureaucratic exercises wasted enormous quantities of money, and vast amounts of time and energy of huge numbers of talented people, to very little benefit. The TQA was especially bad: the cost *per department* was around a third of a million dollars—and the results bore little relationship to the quality of the teaching.

Governments are using taxpayers' money to fund research and teaching. It is of course entirely proper for them to want to ensure that the money is not wasted. However, heavy-handed bureaucratic exercises that themselves waste money are a bad way to achieve this aim. And research flourishes best when the researchers are allowed to use their imaginations, whereas bureaucratic assessment procedures inevitably end up emphasizing "safe" research areas, with a guaranteed—but small—payoff. Instead of guiding research into fruitful areas, officialdom mostly interferes with things that it does not understand. The danger is that, in the long run, this kind of interference will stifle innovation and damage research.

You have Written Your Great Theorem, Now What?

I wish.... Well, what most of us do is go out for a really good meal to celebrate (solving the big problem happens so rarely, you might as well make the most of it). At breakfast the next day you look the gift horse firmly in the teeth and ask: "why did *that* work?" Is there a better way to do that? Can I squeeze a bit more juice out of it? Mathematicians are never satisfied. Oh, and you write it up and publish it. On the web, in a journal, whatever.

And then you move on to another problem.

Where Does One Get an Idea for a Research Paper?

Mine usually come from talking to someone else, or being sent a paper or a preprint that for some reason strikes sparks. The ideas basically just seem to ... happen.

My preference is to keep a lot of problems in mind at the same time. When I get really stuck on one of them, I try a different one for a while. The positive effect of this is that if by chance someone happens to mention an idea that might be relevant, I am more likely to notice. I think that Henri Poincaré got it right when he said that you need a period of conscious work on a problem, after which you should go off and do something different to allow your subconscious to sort the problem out.

Pivotal ideas in my research career have come from (1) A lecture course I attended for the fun of it, (2) A year spent with an American colleague, (3) A phone call from a biologist stimulated by a book review that I had written, (4) A phone call from an engineer who had tried out a technique that I had mentioned in a book on chaos, (5) A casual remark over lunch from a biologist friend, which at first I thought was a misunderstanding, (6) A strange example found by a colleague's postdoc, (7) An e-mail from the authors of a preprint on the arXiv.

Conferences encourage that sort of interaction, especially if they are small and specialized ("Symmetry methods in nonlinear dynamics"), or mix interesting people together with a common aim ("Modeling nature using complex adaptive systems").

Sitting around having coffee in mathematics departments can be an effective source of new ideas. So can a visit to a research centre such as the Santa Fe Institute, MSRI, or the Newton Institute.

A Question You Would Like to Answer or See Answered

Can you infer the qualitative dynamics of a network of coupled dynamical systems—such as whether it is periodic or chaotic—by observing any individual node? The network has to be path-connected, and the result is clearly false without suitable genericity conditions, but there are good reasons to think the statement must be true. However, technical issues make it very hard to prove. If it could be proved, it would be very useful.

Many aspects of mathematics have changed since I started my professional life in 1969 as a temporary lecturer with a one-year position. On the teaching side—which I still keep an eye on, even though I do not give undergraduate lecture courses myself—there is much more paperwork (some justified, some not), and each member of the faculty is now responsible for roughly twice as many students. On the research side, the need to secure research grants and contracts has become incomparably greater: not only do academics do the intellectual work; they also secure much of the money. There have been some major benefits—many more positions for postdoctoral research assistants, for example—but also some disadvantages. The growth of short-term positions at the expense of long-term ones is one of the worst. I am ambivalent about the growing emphasis on commercial applications of university research: such work ought to pay its way, in the long run, but it does no harm if it makes money in the short term. Provided making money does not take over as the sole reason for choosing a particular research topic.

One development, which I applaud (perhaps because for about ten years it was my main responsibility), is the growth of interdisciplinary research in which mathematics is a key component, but not the only one. This is a kind of "new applied mathematics" and it is increasingly becoming the model for the development of the mathematics of the future. However, applications are only one source of mathematical inspiration, and "pure" mathematics deserves support as well, because it is there that many of the most original developments arise.

Forty years ago, you could be a university research mathematician by specializing in your own area of the subject, working largely unassisted, not writing grant proposals, publishing an occasional paper. Today's mathematicians have to be far more versatile, and make contact with industry and business, indeed with the public at large. They have to spend more time on activities that do not contribute directly to research or teaching. Many of them collaborate with people from other areas of science, sometimes even from outside science. This is the way mathematics is now developing, and it brings far more benefits than disadvantages.

I hope that I am not fooling myself by detecting a slight improvement, overall, in the public image of mathematics. Surveys show that mathematicians are respected, that many people think mathematics is useful and deserves support. It may be true that most members of the public have no real idea what mathematics is or what mathematicians do, but the number of people who do understand such things is growing. So overall, I feel that my subject is making good progress and pulling its weight in today's world. There is still much to be done . . . but there always is and always will be.

10

I am a Mathematician

V. S. Varadarajan*

1 Introduction

The title of this piece is exactly the same as Norbert Wiener's famous autobiography [1]. I have chosen it because it summarizes what the editors wanted from me: exactly what is the meaning of this statement, to me, to my colleagues, to the informed scientist, and to the public at large? In the pages to follow I have attempted to present some thoughts in a haphazard and impressionistic manner.

Already in society there is a dichotomy between those who understand to some extent the language of science and have some experience with it, and those who do not. About fifty years ago the British author C. P. Snow discussed this in his 1959 Rede lecture given at Cambridge, U. K., which was later published as a book [2]. The lecture and the book generated intense debate. Snow argued that the widening gap between the humanities and sciences was compromising the ability of the modern society to formulate and solve its problems. He wrote:

> A good many times I have been present at gatherings of people who, by the standards of the traditional culture, are thought highly educated and who have with considerable gusto been expressing their incredulity at the illiteracy of scientists. Once or twice I have been provoked and have asked the company how many of them could describe the Second Law of Thermodynamics, the law of entropy. The response was cold: it was also negative. Yet I was asking something which is about the scientific equivalent of: "Have you read a work of Shakespeare's?"
>
> I now believe that if I had asked an even simpler question such as, What do you mean by mass, or acceleration, which is the scientific equivalent of saying, "Can you read?" not more than one in ten of the highly educated would have felt that I was speaking the same language. So the great edifice of modern physics goes up,

---

*Department of Mathematics, University of California, Los Angeles, CA 90095–1555. email: vsv@math.ucla.edu

and the majority of the cleverest people in the western world have about as much insight into it as their Neolithic ancestors would have had.

The result of this dichotomy that Snow highlighted has been that decisions involving vital matters with a huge scientific component like global warming or stem cell research are being made by people who have no scientific training. In a society which is becoming more and more technologically sophisticated because of the overwhelming influence of computers, this disconnect between science and politics is very unfortunate.

A recent spectacular example of the failure to communicate even to other parts of the scientific community is the case of the American supercollider. The failure of the physicists to make the case for the collider to their own colleagues in other disciplines, and the inability of the politicians to understand the fundamental nature of the project resulted in its cancellation, wasting the gigantic amounts of money, resources, and careers already committed. It was left to the Europeans to continue the project. Part of the reason for the failure was the incompetence of Congress and its inability to understand that this was a project that had truly global implications.

But the gap between those who have some understanding of mathematics and those who do not is even wider and even less easy to bridge. Imagine asking a gathering of intelligent people if they know of noneuclidean geometry, or Fermat's last theorem! Simply put, mathematics is too abstract for most people, and even mathematicians differ widely about what they do or ought to do, and how they should explain themselves to their colleagues and the public at large. Any attempt at a dialogue between mathematicians and the rest of society must start with some explanation of what mathematics attempts to do, what it has done, and how all these things impinge on the way we live and think.

Most mathematicians work in universities or research institutes and so lead very sheltered lives. Their main preoccupation is to discover mathematical theorems and communicate them to others like themselves, and to teach students to become familiar with the mathematical way of thinking so that they can apply it to whatever situation they may be confronted with in their lives. This results, for the majority of the mathematical community (except a subset who are called "applied mathematicians") in a very introspective mode of life in which contact with the real world is minimal, and ultimately, unnecessary. However, we require the outside world for paying our salaries, and for funding what we consider fundamental research. In times of plenty there is no problem about money; but now, with society frittering away its resources in wars and armaments, there is a real shortage of resources, and it has become important to define ourselves more clearly. But while we want to communicate better, I disagree with many of our community who have tried to change their goals to suit what they think will be funded. I firmly believe that we should be the ones who should decide what is best for us, and then make the best possible case for our goals.

Hermann Weyl (1885–1955), one of the greatest mathematicians of the twentieth century, described the difficulties in talking about mathematics to the layman as follows: [3]

> ...Mathematics talks about things which are of no concern at all to man. Mathematics has the inhuman quality of starlight, brilliant and sharp, but cold. But it seems an irony of creation that man's mind knows how to handle things better the farther removed they are from his existence. Thus we are cleverest where knowledge

matters least.... In view of all this: dependence on a long past, other-worldliness, intricacy, and diversity, it seems an almost hopeless task to give a non-esoteric account of what mathematicians have done.

Nevertheless, any attempt to place what we do in the context of the society at large must begin with a description of what appeals to us, what we regard as most valuable, and how we should communicate all this to the public at large. So there is a set of complicated issues involved here and it becomes important to go a little more deeply into them to reach a more coherent picture. This is what I shall attempt now. Since this is not a technical article I will be very informal in most of the things I say. I will concentrate mainly on what makes mathematics so special to its practitioners. From time to time I have allowed myself the liberty of making personal remarks about my development as a mathematician. So in some sense this is a personal statement; more should not be read into it. My aim is to start a discussion, not to present rigid views.

2 The Dual Nature of Mathematics: Some Examples

The most remarkable thing about mathematics is its duality. One can say that mathematics has two faces: an interior one where the criteria for development are purely esthetic, and an exterior one where we have no choice but to follow the path that will lead us to describe Nature, or the world we live in. Of course the experience that goes into the selection of the principles on which we want to pursue a given branch of mathematics might come from our interaction with the physical world, but it need not. But the wonderful and deeply mysterious part of the story of mathematics is that very often precisely those parts that were developed with no thought about applicability to the real world have eventually proved to be the tools for applications, sometimes decades after the discovery of the mathematics itself. Many people, even scientists, have the idea that the mathematician is like the guy in a hardware store; you go to him and state your problem, and he reaches out and gives you the answer. This is not the way things work; mathematics is an organic structure and should be nurtured so that today's research will provide the concepts for understanding and solving tomorrow's problems. This fact should be understood by the people who allocate funds for mathematical research: immediate applicability is too stifling a requirement and will strangle the development of mathematics, hindering the creation of tools for the future.

Harish-Chandra (1923–1983), one of the greatest of contemporary mathematicians, used to refer to the two faces of mathematics as the inner and outer reality. The inner reality is what we have within ourselves. We create mathematical theories based exclusively on their beauty: geometry, number theory, mathematical logic, and so on, are examples of this. Even analysis has to be on this list: the real number system, as conceived by Dedekind (1831–1916) and Cantor (1845–1918), is very far from the phenomenal world. All the mathematical activities that we perform, such as algebra, calculus, trigonometry, solution of equations, the introduction of complex numbers, and so on, are creations within this internal world. Nevertheless, these models of internally created structures are the ones that mimic most accurately the models occurring in Nature, allowing us to describe the external world as an image of the internal world of the mathematician.

How does this phenomenon come about? Why is it that the theory of infinite dimensional spaces, which can have no a priori relation to the three-dimensional world we live in, is precisely what we need to describe the internal structure of the atom and understand the periodic table of elements? To me this is a miracle defying explanation. Perhaps, as we ourselves are a part of the world around us, we may never be able to account for it. Another way to understand it may be as follows. The phenomenal world is very complex and highly interconnected in its parts, and only mathematics is abstract enough to produce models that are sufficiently complex to mirror the external world. I shall begin by discussing two spectacular examples of this feature of life.

Mathematical Description of Atomic Events

Consider for instance the problem of describing the behaviour of an atom, say hydrogen. It is known from experiments that the atom is normally in a stable state (ground state), but if we pump energy into it it will go to an unstable excited state from which it will return to the ground state, emitting the energy back. There are several of these excited levels and to model them accurately and to understand the dynamics of the atom in them it is necessary to separate and differentiate the various excited states by dimension, namely associate a space to the atom where each dimension corresponds to some energy level. In practice we generally have a small number of higher energy levels. However, we cannot always restrict our studies to a single atom: there are other atoms involved, and a little reflection makes it clear that to accommodate all these energy levels we really need an infinite dimensional space! Only the mathematician has experience with these spaces which were created long before any real experiments with atoms were conducted. They were created by David Hilbert (1862–1943) and are therefore called Hilbert spaces. Their geometry is very similar to the geometry of the space we live in—we can speak of distances and angles and other concepts familiar to us from our early experience with geometry, except that they have an infinite number of dimensions, and so their points are described by an infinite number of coordinates. Modern quantum theory needs Hilbert space as an essential mathematical instrument for describing the dynamics of atoms, ions, and elementary particles. Now Hilbert created these spaces at the beginning of the twentieth century, but quantum mechanics was still twenty-five years from being discovered by the physicists!

Measurements of energy levels and states of atoms are done by spectroscopy, and the advances in spectroscopic technology in the early days of the twentieth century were responsible for the discovery of the physicists that the classical mechanics of Newton and Maxwell was inadequate to produce models in which the behaviour of atoms could be discussed, available information organized, and new phenomena predicted. Conventional representations of classical mechanics and electrodynamics had to be abandoned, and one had to use the geometry of the Hilbert spaces to represent faithfully what is happening inside an atom. The impossibility of measurements of position as well as velocity of an atomic particle, due to the fact that the objects to be measured are of the same size as the measuring instruments, thus erasing the distinction between the observer and the observed world, gives rise to the fact that all quantum mechanical statements are ultimately statistical. No models other than those coming from Hilbert space can accurately depict and predict these measurement uncertainties in quantum theory.

I have glossed over several aspects of this model because I did not wish to make the discussion too difficult. In actuality, the vectors in the Hilbert space of quantum theory are complex, not real, and although they represent the states of the system, two vectors that are proportional represent the same state. The representation of states by vectors up to proportionality is called the superposition principle by physicists. Only with such vector representations are we able to picture in a coherent way the wave-particle duality of matter, namely the profound experimental fact that all material particles including light have both wave and particle properties. This is called the complementarity principle.

Probability Theory

Let me now take a second example, namely, probability theory. It was originally born in an exchange of letters between two French mathematicians, Blaise Pascal (1623–1662) and Pierre de Fermat (1601–1665), to explain some observations made by one of their friends, the Chevalier de Méré who ran a gambling casino. The specific problem was the following: in the case of two players who want to finish a game early, how to divide the stakes fairly, based on the chance each has of winning the game from that point. Eventually this discussion led to the concept of expected values and to the creation of the theory of probability. With the tools created by the above two and by Jacob Bernoulli (1654–1705), probability theory developed rapidly and was able to provide a clear mathematical foundation for all practical problems in which statistics played a role.

In the twentieth century Norbert Wiener (1894–1964) advanced the theory in a fundamental way by showing how probabilities can be associated to the motion of very small bodies immersed in a liquid medium under the bombardment of the molecules of the medium. This is the so-called Brownian Motion, first discovered by an English botanist Robert Brown (1773–1858). This motion was studied experimentally by the physicist Jean-Baptiste Perrin (1870–1942), who won the Nobel prize for his studies, and who suggested that the mathematical modeling of it might be worthwhile. Because the motion is caused by the unpredictable actions of billions of molecules it is clear that any such model has to be probabilistic. However the construction of a probabilistic model for this motion led to profound mathematical questions. The point is that, if you follow the motion, you have to plot the positions of the particle at various instants of time (not an easy matter, and Perrin had to devise ingenious techniques for his experimental work), so that in principle, each observation is a path. Thus Wiener's problem was how to define probabilities in a space of paths. The space of paths is very different from the normal spaces we encounter in that it is also of infinite dimension, like the Hilbert spaces of quantum theory discussed above. Perrin had observed that the paths that he plotted reminded him of the continuous functions which had no derivatives anywhere, because, although the paths did not jump, they were changing directions in such a way that no tangent could be associated to them at any point. Wiener not only constructed the way to compute probabilities for the brownian paths but also showed that Perrin's observation admitted a proof in his model.

Wiener's discovery is a watershed event in the theory of probability and moved it, if not at once, but certainly in a few decades, to a position of central importance in modern mathematics. Nowadays it is used in predicting the movement of stocks in the market and other sophisticated aspects of finance. Yet, no one, least of all Wiener himself, could have foreseen the modern uses of probability theory.

In the 1940s, a young physics graduate student at Princeton University, Richard Feynman (1918–1988), discovered a completely novel and startling way to formulate quantum mechanics by using probabilities in path space, in a manner similar to, but more complicated than, that of Wiener. His method consisted in associating a complex probabilty amplitude to each path that takes a particle (say) from one point to another, and adding up these complex amplitudes to compute the final probability as the square of the absolute value of the total amplitude. To make this mathematically rigorous was (and remains) extremely difficult; in some cases it has been accomplished by Mark Kac (1914–1984), who showed that one can use Wiener's work to compute the probabilities à la Feynman!

I was a graduate student in the Indian Statistical Institute at Calcutta, India when I came across Feynman's article written for probabilists, explaining his ideas for computing probabilities for atomic events, concentrating specifically on the famous double-slit experiment. To this day I remember how stunning this was to me. It kindled a deep desire within me to understand how quantum theory is formulated mathematically. That it could be based on Wiener's theory, which I knew very well was an added incentive. Nowadays all of physics is being formulated in this framework of ideas, and Feynman's technique is usually referred to as the sum-over-histories method.

The two examples given above are among many and they all point to the same conclusion: one must not suffocate the development of new ideas and directions of research in mathematics because one does not know where the new demand will come from.

In retrospect, I think the fact that my postgraduate training was at a place where both theory and applications were treated with equal respect, was a blessing in disguise. It developed a sense of real appreciation in me for people who were trying to use mathematics to explain the world around them, not only in the traditional sciences like physics and chemistry, but also in less conventional (at that time) areas like statistics, biometrics, econometrics, and psychometry. Prasanta Chandra Mahalanobis (1893–1972) who founded the Institute, and Calyampudi Radhakrishna Rao (1920–) who directed the research wing of the Institute, were true visionaries; they had a global conception of science in general and mathematics in particular, and encouraged our participation in a wide variety of activities, both theoretical and applied. We had visitors who were experts in the most diverse fields, and our interaction with them was crucial to the growth and development, not only of myself but of all the graduate students of the Institute. I remember a lecture by Bogolyubov on the use of distributions in quantum field theory, lectures by Norbert Wiener on prediction theory, lectures by economists Galbraith and Kosygin who would later become powerful political figures, several series of lectures by J. B. S. Haldane on biometric aspects of statistics, lectures by the mathematician Linnik on sufficient statistics, and by Mahalanobis himself on the graphical analysis of statistical data.

It was during a later stay at the Institute that I worked with Professor Radhakrishna Rao on the problem of statistical discrimination between two alternatives, when the number of variables is so large that the problem really becomes one in a space of infinitely many dimensions. This collaboration led to my only(!) applied papers. Years later, when I was at UCLA I met Charles Baker, an engineering student, who told me that the results of this paper were often used in sonar detection of submarines! This was very remarkable for me; at the time when the paper was conceived and worked over, there was really no idea of specific applications, and the problem was posed purely as the study of equivalence and orthogonality of two Gaussian probability measures in a real Hilbert space, and in case of equivalence, to the calculation of the mutual Radon-Nikodym derivatives (that there are only these two alternatives is a result that goes back in an essential form to Shizuo Kakutani (1911–2004), perhaps even to Wiener's work; it has been proved by many people since, and repeatedly!).

3 Geometry as a Case History of the Duality in Mathematics

This duality discussed briefly above, which is at the very center of my own personal motivation to study mathematics, deserves further elaboration. I have therefore chosen to illustrate it further with an account of the historical development of geometry, because many people are familiar with geometry and one does not need very sophisticated ideas to explain the evolution of the subject. One can clearly see from this account that the great ideas were driven by purely internal questions, but the end product became surprisingly flexible and therefore applicable to all sorts of situations, from the quantum structure of the atom to the structure of the entire cosmos, including black holes.

The origins of geometrical thought go back to very old times but for almost all of us the first introduction to it must have been when we studied Euclid's geometry. Euclid was a Greek mathematician who lived and worked around 300 B. C. and wrote what is arguably the most famous mathematical treatise of all time, *The Elements*, in 13 volumes, describing the state of the knowledge of geometry at that time. Although he treated many parts of mathematics in *The Elements*, the core of his work was devoted to the geometry of the plane. This plane is now called the Euclidean Plane. It is of infinite extent in all directions. It has points and lines, the lines are of infinite extent, and a crucial concept, that of parallel lines, is conceptually permissible precisely because of this infinite extendability of the lines. The real essence of Euclid's geometry lies in the fact that only by assuming that lines extend infinitely in both directions and further that there are parallel lines, that can one construct a beautiful theory.

Euclid made a sharp distinction between axioms (as we call them nowadays) and propositions. Axioms are, naively speaking, what we are willing to assume without proofs, and propositions are the ones that can be logically deduced from the axioms. In this way he built a wonderful structure that dominated scientific and philosophical thought for a very long time. In fact so convincing was his structure and so masterful his development of that structure, that many thinkers (Kant, for example) seriously believed that Euclid's geometry was the only one possible, and that therefore it described precisely the geometry of the physical world we live in. Even a supreme mathematician such as Gauss (1777–1855), who knew the real nature of these questions and who had discovered an alternative to Euclidean geometry, did not publish his findings because he feared the "outcry of the Boeotians." It was left to Johann Bolyai (1802–1860) and Nikolai Ivanovitch Lobachevsky (1793–1856) to publish their discovery of non-euclidean geometry (around 1826).

Euclid himself planted the first seeds of doubt on his geometry when he introduced the axiom of parallels. In the form we give to it nowadays, it says that if a line and a point not on it are given, there is exactly one line through that point that does not meet the given line, i.e., is parallel to it. Euclid was aware of the highly counterintuitive nature of this axiom and so, perhaps in the hope that eventually a way could be found to eliminate this axiom, began his development of geometrical propositions where he did not use the parallel axiom. In some sense one may say that Euclid is the first non-euclidean geometer! But very soon he found that it was impossible to proceed without this axiom. For one of the most central facts of geometry, namely, the assertion that the sum of the three angles of a triangle is two right angles, he needed the parallel axiom.

Euclid's ambivalence about the parallel axiom did not escape the notice of his successors. For many centuries they vainly tried to deduce it from the other axioms but did not succeed. Eventually Bolyai and Lobachevsky discovered non-euclidean geometry; Gauss's discoveries became known only from his posthumous papers.

However, as a newcomer, non-euclidean geometry had to pass severe tests to qualify as a legitimate competitor to Euclid's geometry which had survived two thousand years. The question arose whether the geometry of Bolyai and Lobachevsky was internally consistent. When this question was formulated, it became clear that it could have been asked also of Euclid's geometry. It was Felix Klein (1849–1925) who settled this issue to some extent by constructing a Euclidean model for non-euclidean geometry. Klein's beautiful construction showed that, if we assume the consistency of Euclid's geometry, then the consistency of the geometry of Bolyai and Lobachevsky must follow as a consequence.

Klein started with the unit circle U in the Euclidean plane. Only the points inside U are the points of the non-Euclidean (n. e.) geometry. The lines of the n. e. geometry are the parts of the euclidean lines within U. A pair of n. e. lines that do not meet inside U are thus parallel. One can define distances and angles in such a way that all axioms of Euclid hold, except the parallel axiom. Indeed, given a n. e. line and a n. e. point not on it, there is an infinite number of lines through the point parallel to the given line (as an exercise, the reader should draw the obvious figure).

The concept of parallel lines was at the center of another great advance in geometry, namely the subject of projective geometry, discovered by Gérard Desargues (1591–1661) among others and systematized by Jean-Victor Poncelet (1788–1867). Projective geometry arose from the observation that, by projecting from various points in space, one can convert lines that are parallel in one plane into lines in another plane that meet, and vice versa. One might say that the process of projection makes a line or a point go to infinity, i.e., become a vanishing line or a vanishing point, and conversely, make a point or line at infinity reappear in the finite part of a plane. This discovery is even older and goes back to the painters who discovered the concept of perspective in art. Of course, by allowing such projections from one plane to another, distances would not be preserved—which is otherwise also obvious from the fact that finite points are sent to infinity and vice versa. So Euclid's geometry is extended by eliminating parallel lines; a new line is introduced, called the line at infinity, whose points are the points at infinity, and parallel lines are assumed to meet at points of this line. This new geometry is called projective geometry; any two distinct lines in this geometry always meet at a point, there are no distances and angles, and therefore only properties of incidence are studied. There are no circles, but it makes sense to speak of conics. This geometry is also very beautiful. Pascal's hexagrammum mysticum is a configuration in projective geometry. One starts with a hexagon inscribed in a conic and finds the intersections of the three pairs of opposite edges (with respect to a given ordering of the six vertices); then the three points of intersection lie on a line, called the Pascal line. One obtains many Pascal lines by changing the order of the vertices of the hexagon, actually 60 of them. The 60 Pascal lines are concurrent, 3 at a time, in 20 Steiner points, which, 4 at a time, lie on 15 Plücker lines, and so on. The entire configuration is a fantastic one and is a prime example of the beauty of projective geometry.

Projective geometry has many configurations, namely, assemblages of points and lines with a variety of properties of intersections. The first configurations were constructed by Desargues, although there is an earlier, very much more ancient one, due to Pappus of Alexandria (c 290 C.E.–c 350 C.E.). The Desargues's configuration is about two triangles in perspective, and the theorem it refers to is that the points of intersections of the three corresponding pairs of sides of the triangles are on a line. The Pappus configuration is a degenerate version of the Desargues configuration.

Euclid's geometry can be recovered from projective geometry by singling out a line and calling it the line at infinity and defining two lines as parallel if they meet on this line. Poncelet had discovered that there are two special points on the line at infinity through which all circles must pass, the so-called circular points at infinity, so that the theory of circles, and more generally, distances and angles may be defined using projective concepts.

In the seventeenth century René Descartes (1596–1650), one of the most influential figures in the scientific revolution and the author of the famous words *cogito, ergo sum* ("I think, therefore I am"), discovered a way to represent geometrical figures and loci by algebraic equations, thus converting geometry into algebra. This was the birth of algebraic geometry and it had a profound effect on the development of mathematics. Geometrical points were described by their coordinates and lines and other loci like the circle and ellipse by the equations that their coordinates satisfied. The distance between two points was described by a quadratic function of the coordinates which encoded the theorem of Pythagoras. This merging of geometry with algebra might be regarded as the most important discovery of the seventeenth and eighteenth centuries. The foundations of algebra and the number system however were not properly understood in those early days and people appeared to think that synthetic geometry, as it was called to distinguish it from Descartes's analytic geometry, was somehow more general. It was only after the development of algebraic geometry that this issue became clear.

By very abstract methods, the projective geometers succeeded in establishing coordinates in any projective geometry. The validity of Desargues' theorem, automatic in dimension three but not in dimension two, is needed to establish that coordinates can be constructed for the geometry from a division ring; the validity of the theorem of Pappus is needed to assert that this division ring is a field, i.e., is commutative.

These ideas play a surprising role in the foundations of quantum geometry [4]. Indeed, according to John Von Neumann (1903–1957), the experimental propositions of quantum theory are arranged as a projective geometry, the principle of superposition corresponding to collinearity. Thus the field of coordinates is a deep intrinsic invariant of the system. In quantum theory this field is assumed to be the complex number field, to allow for phenomena like charge conjugation.

These discoveries bolstered the belief that, although Euclid's geometry was not unique, somehow the axioms limited the choice of possible geometries. However, this could not be proved, and there remained the possibility of other types of geometries being discovered. It was Gauss and Riemann (1826–1866) who took the whole subject to a new level, and discovered differential geometry. First Gauss discussed the geometry of surfaces and discovered the notion of curvature. Then Riemann considered spaces of arbitrary dimension and built all the geometries that are Euclidean at infinitesimal distances, but diverge from it over finite parts of the world. He introduced the curvature of such a space and showed that the vanishing of this curvature is the condition that space is Euclidean, not only infinitesimally, but over finite parts as well. The geometries of Euclid resp. Bolyai-Lobachevsky now emerged as the ones where the curvature is everywhere 0 (resp. constant, resp. < 0). The theorem that the sum of the three angles of a traingle is two right angles is precisely the vanishing of the curvature. The vanishing of curvature is called flatness for obvious reasons.

The square of the distance in Riemann's geometry between points infinitesimally near to each other is a quadratic function of the differences of the coordinates. Riemann allowed also the possibility that the fourth power of the distance is a polynomial of the fourth degree. Subsequently, with the development of what are now called Finsler geometries, due to Paul Finsler (1894–1970), one can view Riemann's construction as a part of a more general scheme. It would be interesting to understand the fourth power case of Riemann a little better.

I have referred to axioms and propositions only in a naive way. At the end of the nineteenth and the beginning of the twentieth centuries a searching reexamination of Euclid's geometry and its axiomatic foundations was undertaken by Hilbert. Using the fact that coordinates can be introduced one can reduce the consistency of geometry to that of arithmetic. Surprisingly this proved elusive to Hilbert and his disciples including John Von Neumann. The whole subject took a dramatic turn when Kurt Gödel (1906–1978) showed that any axiomatic system such as arithmetic must admit propositions that are true but not provable within the system. For a glimpse into Hilbert's work, see reference [5].

The entire development of geometry, culminating in Riemann's ideas, stood in splendid isolation as a monolithic tribute to the power of speculative thought until Einstein (1879–1955) used Riemann's themes and showed that the geometry of space-time is non-euclidean and it is this non-euclidean aspect that gives the true explanation for the universal phenomenon of gravitation. This was a stunning achievement for several reasons. First, space was replaced by space-time, a combination of space and time. Before Einstein, for instance with Newton, space and time were absolute entities, appearing exactly in the same way to all observers. Einstein demolished this idea, showing convincingly that time flows differently for each observer because there is an absolute limit to the speed of propagation of signals, namely the velocity of light in a vacuum. Because of this only space-time has an absolute significance for all observers. Space-time became a geometrical object like Riemann's but with a metric that was not Pythagorean but of signature (3, 1), namely that the infinitesimal distance is no longer

$$dx_0^2 + dx_1^2 + dx_2^2 + dx_3^2$$

but

$$-dx_0^2 + dx_1^2 + dx_2^2 + dx_3^2.$$

Using fully Riemann's apparatus Einstein showed that, even in the absence of matter, the curvature of space-time is not in general 0 but obeys certain equations which explain gravitation. Among the most spectacular predictions of Einstein's theory is the bending of light by gravitational fields, verified by observations on an annular solar eclipse in 1919. Einstein's theory also provides the foundation for the study of black holes, a subject of intense interest to cosmologists.

If this were all it would be dramatic enough. But, in the 1970s, physicists, searching for a way to study the dynamics of elementary particles that treats both bosons (like photons) and fermions (like electrons) on the same footing, discovered an extension of geometry where the coordinates are the usual ones supplemented by what are called Grassmann coordinates. Such a space is a supergeometric object. Supergeometry is an extension of usual geometry which is highly non-trivial and is a reflection of the structure of space-time at distances that are extremely small even by the standards of elementary particle theory. It is quite likely that experiments done with the new super collider being readied at CERN will confirm that the world is supergeometric at ultrasmall distances and times. The Grassmann coordinates cannot be seen or measured, but they affect the dynamics of particles and fields in essential ways. For a discussion of the evolution of geometry leading up to supergeometry (see the reference [6]).

To me, the discovery of supergeometry is a fantastic illustration of the interaction of mathematics and physics. Modern elementary particle physics has many unsolved issues and it appears that only with the deepest interaction with mathematics can one hope to make some theoretical progress on these, mainly because the energies at which the new issues can be experimentally resolved is out of reach for a very long while into the future.

4 Computers

I have postponed to the end any discussion of computers simply because I am not competent to say anything really significant. The last quarter of the twentieth century should surely be described as the start of the age of the computer. The computer revolution has changed the very core of the lives of everyone on this planet in ways that could not have been imagined even ten years ago. But the origins of this revolution go back to the work of Alan Turing (1912–1954) on the structure of algorithms ("Turing machines"), and the work of John von Neumann on the structure of the brain, theory of cellular automata, and so on. Much of this work was highly theoretical and was in no way propelled by any possibility of future applications, although advances in scientific computing owed their impetus to the Manhattan Project and problems of numerical weather prediction. The computer world has even entered the world of the mathematician, by providing the researcher fast methods of computation that would have been unthinkable just a few years ago. For instance, a substantial part of the famous four color problem was done entirely using computers [7].

5 How to make Ourselves Heard Better

I have sketched some parts of the story of how mathematicians tend to view their subject. It is clear that even with the greatest skill in exposition and good will, this story is simply too difficult for the layman, or even the scientifically cultured person, to absorb and understand. What we, as a mathematical community, need to do is to try to get this story across by expositions that are simple, not very technical, and keep the spirit of the development intact. This should be done at the level of high school as well as the university. The high school texts in mathematics are really moribund and do not convey the excitement and mystery of the basic ideas. The texts in the university are even worse, if such a thing can be imagined. Calculus, for example, is presented as a collection of recipes, as in a cook book. A student taking calculus will have, typically no idea that he is learning about what is arguably one of the greatest revolutions in scientific thought. This situation has to change, and the student should be exposed to the basic ideas, if not in text books, at least in seminar courses designed specifically for such a purpose.

This is a problem that is highly political and requires active involvement by the mathematician in decisions that affect the community. Unfortunately this needs the mathematician to come down from the safe ivory tower of ideas and introspection and fight it out in the real world where agendas abound and motives are extremely complex. I have no ideas to offer as to how this can be achieved.

References

[1] Norbert Wiener, *I am a Mathematician*, Doubleday, 1956.

[2] C. P. Snow, The Two Cultures and the Scientific Revolution (quotation in text taken from The Wikipedia).

[3] Hermann Weyl, A half-century of mathematics, *Amer. Math. Monthly*, 58(1951), 523–553. Ges. Abh. Band IV, pp. 464–494, Springer-Verlag, 1968.

[4] V. S. Varadarajan, Geometry of Quantum Theory, Second Edition, Springer-Verlag, 1985, 2007.

[5] Hermann Weyl, David Hilbert and his mathematical work, Bull. Amer. Math. Soc., 50(1944), 612–654. Ges. Abh. Band IV, pp. 132–172, Springer-Verlag, 1968.

[6] V. S. Varadarajan, Supersymmetry for mathematicians:an introduction, Courant Lecture Notes in Mathematics, 11, New York University, Courant Institute of Mathematical Sciences, New York; Amer. Math. Soc., Providence, 2004.

[7] Robin Wilson, Four Colours Suffice: How the Map Problem Was Solved, Princeton University Press, princeton, New Jersey, 2004.

Part II

On Becoming a Mathematician

Foreword to
On Becoming a Mathematician

If one wants to become a mathematician, then one is faced with the prospect of a goodly number of years of schooling. Usually this culminates with a Masters Degree or a Ph.D. It is just a fact of life that most of that schooling concentrates on the *learning of mathematics* and, in the case of the Ph.D., on the *discovery of new mathematics*. Perhaps an insufficient amount of this time and effort is devoted to ideas about teaching.

It stands to reason, then, that a mathematician writing about "On Becoming a Mathematician" will rely on his/her own personal experience in graduate school, and as an Assistant Professor, to describe what this maturation process consists of. And a key feature of that genesis will be research. When we think about the path that we followed to become a mathematician, we think about papers that we struggled to write, journals that we worked with to get our work published, grants that we applied for, conference invitations that we hoped to receive, and promotions that we sought. Of course teaching was also a big part of our lives, and we worked hard to become excellent teachers. We would be disingenuous to suggest that teaching is not of utmost significance. And we are proud of our teaching. But, when someone asks me what is involved in becoming a mathematician, it is the mathematics that looms large.

The essays in this part of the book reflect the philosophy described in the preceding two paragraphs. They certainly talk about teaching and communicating, and well they should. But the emphasis is on the subject matter itself. The reader should balance these pieces against those appearing in the "Who are Mathematicians?" part of the book in order to get a complete picture of what this work is about, and what message it is endeavoring to convey. Everyone who wrote for this volume loves mathematics, and wants to share that enthusiasm and passion.

—Peter Casazza, Columbia, Missouri
—Steven G. Krantz, St. Louis, Missouri
—Randi D. Ruden, University City, Missouri

11

Mathematics and Teaching

Hyman Bass

Induction

I was not born to be a mathematician. Like many, I was drawn to mathematics by great teaching. Not that I was encouraged or mentored by supportive and caring teachers; such was not the case. It was instead that I had as teachers some remarkable mathematicians who made the highest expression of mathematical thinking visible, and available to be appreciated. This was like listening to fine music, with all of its beauty, charm, and sometimes magical surprise. Though not a musician, I felt that this practice of mathematical thinking was something I could pursue with great pleasure, and capably so even if not as a virtuoso. And I had the good fortune to be in a time and place where such pursuits were comfortably encouraged.

The watershed event for me was my freshman (honors) calculus course at Princeton. The course was directed by Emil Artin, with his graduate students John Tate and Serge Lang among its teaching assistants. It was essentially a Landau-style course in real analysis (i.e., one taught rigorously from first principles). Several notable mathematics research careers were launched by that course. Amid this cohort of brilliant students, I hardly entertained ideas of an illustrious mathematical future, but I reveled in this ambience of *beautiful thinking*, and I could think of nothing more satisfying than to remain a part of that world. It was only some fifteen years later that I came to realize that this had not been a more or less standard freshman calculus course.

Certain mathematical dispositions that were sown in that course remain with me to this day, and influence both my research and my teaching. First is the paramount importance of proofs, as the defining source of mathematical truth. A theorem is a distilled product of a proof, but the proof is a mine from which much more may often be profitably extracted. Proof analysis may show that the argument in fact proves much more than the theorem statement captures. Certain hypotheses may have been not, or only weakly, used, and so a stronger conclusion might be drawn from the same argument. Two proofs may be observed to be structurally similar, and so the two theorems can be seen to be special cases of a more unifying claim. The most agreeable proofs explain, rather than just establish, truth. And the logical narrative clearly distinguishes the illuminating turn from technical routine.

Artin himself once reflected on teaching in a review published in 1953.

> We all believe that mathematics is an art. The author of a book, the lecturer in a classroom tries to convey the structural beauty of mathematics to his readers, to his listeners. In this attempt, he must always fail. Mathematics is logical to be sure, each conclusion is drawn from previously derived statements. Yet the whole of it, the real piece of art, is not linear; worse than that, its perception should be instantaneous. We have all experienced on some rare occasion the feeling of elation in realising that we have enabled our listeners to see at a moment's glance the whole architecture and all its ramifications.

Two things of a more social character about mathematics also impressed me. First, the standards for competent and valid mathematical work appeared to be clear, objective, and (so I thought at the time) culturally neutral. The norms of mathematical rigor were for the most part universal and shared across the international mathematics community. Of course mathematical correctness is not the whole story; there is also the question of the interest and significance of a piece of mathematical work. On this score matters of taste and aesthetics come into play, but there is still, compared with other fields, a remarkable degree of consensus among mathematicians about such judgments. One circumstance that readily confers validation is a rigorous solution to a problem with high pedigree, meaning that it was posed long ago, and has so far defied the efforts of several recognized mathematicians.

A social expression of this culture of mathematical norms stood out to me. Success in mathematics was independent of outward trappings—physical or social—of the individual. This was in striking contrast with almost every other domain of human endeavor. People, for reasons of physical appearance, affect, or personality, were often less favored in non-mathematical contexts. But, provided that they met mathematical norms, such individuals would be embraced by the mathematics community. At least so it seemed to me, and this was a feature of the mathematical world that greatly appealed to me. I have since learned that, unfortunately, many mathematicians individually compromised this cultural neutrality, allowing prejudice to discourage the entry of women and other culturally defined groups into the field.

One effect of this intellectual indifference to social norms in mathematics is that a number of accomplished mathematicians do not present the socially favored images of appearance and/or personality, and so the field is sometimes caricatured a being one of brilliant but socially maladroit and quirky individuals. On the contrary, it is a field with the full range of personality types, and it is distinctive instead for its lack of the kind of exclusion based on personality or physical appearance that infects most other domains of human performance. Perhaps the intellectual elitism and sense of aristocracy common to many mathematicians is a counterpoint to this social egalitarianism.

The second social aspect of mathematics that stood out to me concerned communication about mathematics to non-experts. Throughout my student years, undergraduate and graduate, I was amazed and excited by the new horizons being opened up. I enthusiastically tried to communicate some of this excitement to my non-mathematical friends, from whom I had eagerly learned so much about their own studies. These efforts were increasingly frustrated, despite my efforts to put matters in analytically elementary terms. I think perhaps that I had already become too much of a mathematical formalist, and considered the formal rendering of the ideas an important part of the message. This rendering was often inaccessible to my friends, despite my enthusiasm. As my research career entered more abstract theoretical domains I gradually, and with disappointment, retreated somewhat from efforts to talk to others about what I did as a mathematician. Samuel Eilenberg, my mentor when I first joined the Columbia University faculty, once said something to the effect that,

> *Mathematics is a performance art, but one whose only audience is fellow performers.*

I remain to this day deeply interested in this communication problem, and I have admired and profited from the growing numbers of authors who have found the language, representations, styles, and narratives with which to communicate the nature of mathematics, its ideas and its practices. Also, my current studies of mathematics teaching have reopened this question, but now in a somewhat different context. For twelve years of schooling, mathematics has a captive audience of young minds with a natural mathematical curiosity too often squandered. And these children's future teachers are students in the mathematics courses we teach.

Mathematical Truth and Proof

Each discipline has its notions of truth, norms for the nature and forms of allowable evidence, and warrants for claims. Mathematics has one of the oldest, most highly evolved and well-articulated systems for certifying knowledge—deductive proof—dating from ancient Greece, and eventually fully formalized in the twentieth century. There may be philosophical arguments about allowable rules of inference and about how generous an axiom base to admit, and there may be practical as well as philosophical issues about the production and verification of highly complex and lengthy, perhaps partly machine-executed, "proofs." But the underlying conceptual construct of (formal) proof is not seriously thrown into question by such productions, only whether some social or artifactual construction can be considered to legitimately support or constitute a proof.

Mathematical work generally progresses through a trajectory that I would describe as,

$$\text{Exploration} \rightarrow \text{discovery} \rightarrow \text{conjecture} \rightarrow \text{proof} \rightarrow \text{certification}$$

In my work with Deborah Ball (2000, 2003) we have described the first three phases as involving *reasoning of inquiry*, and the last two as *reasoning of justification*. The former is common to all fields of science. The latter has only a faint presence in mathematics education, even though it is the distinguishing characteristic of mathematics as a discipline.

Deductive proof accounts for a fundamental contrast between mathematics and the scientific disciplines; they honor very different epistemological gods. Mathematical knowledge tends much more to be cumulative. New mathematics builds on, but does not discard, what came before. The mathematical literature is extraordinarily stable and reliable. In science, in contrast, new observations or discoveries can invalidate previous models, which then lose their scientific currency. The contrast is sharpest in theoretical physics, which historically has been the science most closely allied with the development of mathematics.

I. M. Singer is said to have once compared the theoretical physics literature to a blackboard that must be periodically erased. Some theoretical physicists—Richard Feynman, for instance—enjoyed chiding the mathematicians' fastidiousness about rigorous proofs. For the physicist, if a mathematical argument is not rigorously sound but nonetheless leads to predictions that are in excellent conformity with experimental observation, then the physicist considers the claim validated by nature, if not by mathematical logic; nature is the appropriate authority. The physicist P. W. Anderson once remarked, "We are talking here about theoretical physics, and therefore of course mathematical rigor is irrelevant."

On the other hand, some mathematicians have shown a corresponding disdain for this freewheeling approach of the theoretical physicists. The mathematician E. J. McShane once likened the reasoning in a "physical argument" to that of "the man who could trace his ancestry to William the Conqueror, with only two gaps."

Even some mathematicians eschew heavy emphasis on rigorous proof, in favor of more intuitive and heuristic thinking, and of the role of mathematics to help explain the world, in useful or illuminating ways. In general they do not necessarily scorn rigorous proof, only consign it to a faintly heeded intellectual super-ego. But I would venture nonetheless that, for most research mathematicians, the notion of proof, and its quest, is at least tacitly central to their thinking and their practice as mathematicians. And this would apply even to mathematicians who, like Bill Thurston, view what mathematicians do as not so much the production of proofs, but as "advancing mathematical understanding" (Thurston, 1992). It would be hard to find anyone with the kinds of mathematical understanding and function of which Thurston speaks who has not already assimilated the nature and significance of mathematical proof.

At the same time, the writing of rigorous mathematical proofs is not the work that mathematicians actually do, for the most part, or what they most cherish and celebrate. Such tributes are conferred instead on acts of creativity, of deep intuitive discovery, of insightful analysis and synthesis. André Weil (1971) said,

> *If logic is the hygiene of the mathematician, it is not his (sic) source of food; the great problems furnish the daily bread on which he (sic) thrives.*

Vladimir Arnold offered an even more derisory characterization:

> *Proofs are to mathematics what spelling (or even calligraphy) is to poetry. Mathematical works do consist of proofs, just as poems do consist of characters.*

But it is proof that finally gives mathematical achievements their pedigree.

The Proof vs. Proving Paradox

Saying that a mathematical claim is true means, for a mathematician, that there exists a proof of it. Strictly speaking, this is a theoretical concept, independent of any physical artifact, and therefore also of any human agency. But *proving* is an undeniably human activity, and so susceptible to human fallibility. It is an act of producing conviction about the truth of something. A mathematician, in proving a mathematical claim, will typically produce a manuscript purporting to represent a mathematical proof, and exhibit it for critical examination by expert peers (certification). But, as Lakatos (1976) has vividly described, this process of proof certification can be errant and uneven, though ultimately robust.

The rules for proof construction are sufficiently exact that the checking of a proof should, in principle, be a straightforward and unambiguous procedure. However, formal mathematical proofs are ponderous and unwieldy constructs. For mathematical claims of substantial complexity, mathematicians virtually never produce complete formal proofs. Indeed, requiring that they do so would cause the whole enterprise to grind to a halt. The resulting license in the practices of certifying mathematical knowledge has caused some (nonmathematical) observers to conclude that mathematical truth is just another kind of social negotiation, and so, unworthy of its prideful claims of objective certainty.

I think that this reasoning is based on a misunderstanding of what mathematicians are doing when they claim to be proving something. Specifically, I suggest that:

> *Proving a claim is, for a mathematician, an act of producing, for an audience of peer experts, an argument to convince them that a proof of the claim exists.*

Two things are important to note here. First, that implicit in this description is the understanding that the peer experts possess the conceptual knowledge of the nature and significance of mathematical proof. Second, the notion of "conviction" here is operationalized to mean that:

> *The convinced listener feels empowered by the argument, given sufficient time, resources, and incentive, to actually construct a formal proof.*

Of course, in this time of computer-aided proofs, some of this certification must be transferred to establishing the reliability of proof-checking software.

In my work with Deborah Ball (2000, 2003), we have found this perspective helpful in studying the development of what might reasonably be called mathematical proving in the early grades. I say more about this below.

Compression and Abstraction

Mathematical knowledge is, as I mentioned above, cumulative (nothing is discarded), and also hierarchical. What saves it then from sinking under the weight of its own relentless growth, into some dense impenetrable massive network of ideas? It is rescued from this by a distinctive feature of mathematics that some have called *compression*. This is a process by which certain fundamental mathematical concepts or structures are characterized and named, and so cognitively rescaled so that they become, for the expertly initiated, as mentally manipulable as are counting numbers for a child. Think for example what a complex of ideas (and mathematical history) is packed into an expression like "complex Lie group," uttered fluently among mathematicians. Think of the years of schooling needed to invest that expression with precise meaning.

A typical form of compression arises from the unification of diverse phenomena as special cases of a single construct (for example groups, topological spaces, Hilbert spaces, measure spaces, categories and functors, etc.) about which enough of substance can be said in general to constitute a useful unifying theory. This leads to another salient feature of mathematics: *abstraction*. Most mathematics has its roots in science, and so ultimately in the "real world." But mathematics, even that contrived to solve real world problems, *naturally* generates its own problems, and so the process continues with these, leading to successive stages of unification through abstraction, until the mathematics may be several degrees removed from its empirical origins. It is a happy miracle that this process of pursuing *natural mathematical questions* repeatedly reconnects with empirical reality, in unexpected and unplanned ways. This is the "unreasonable effectiveness" of which Wigner wrote (1960) and of which Varadarajan (2010) writes eloquently in this collection of essays.

Abstraction is often thought to separate mathematics from science. But even among mathematicians there are different professed dispositions toward abstraction. Some mathematicians protest that they like to keep things "concrete," but a bit of reflection on what they consider to be concrete will show it to be far from such for an earlier generation. Indeed I noticed, while a graduate student, that the extent to which a mathematical idea was considered abstract seemed more a measure of the mathematician's age than of the cognitive nature of the idea. As new ideas become assimilated into courses of instruction, they become the daily bread and butter of initiates all the while remaining novel and exotic to many of their elders.

While compression is an essential instrument for the ecological survival of mathematics, its very virtue presents a serious obstacle for the teaching of mathematics. The knowledgeable and skillful mathematician has assimilated and internalized years of successive compression, streamlining of ideas, and habits of mind. But a young learner of mathematics still lives and thinks in a "mathematically decompressed" world, one that has become hard for the mathematically trained person to imagine, much less remember. This presents a special challenge to teachers of mathematics to children. And, interestingly, it requires a special kind of knowledge of mathematics itself, which is neither easy for nor common among otherwise mathematically knowledgeable adults. (See Ball et al., 2005, 2008.)

Teaching Mathematics

Among the questions to which our editors invited us to direct attention was, "How are we research mathematicians viewed by others?" For one community, (school) mathematics teachers and education researchers, I have some firsthand knowledge of this, after more than a decade of interdisciplinary work in mathematics education. And what I have learned I find both interesting, and important for mathematicians to understand. I began this essay with an account of how great university teaching drew me into mathematics. Now, in my close study of elementary mathematics teaching, I have a changed vision of what teaching entails.

Of course many mathematicians, like mathematics educators, are seriously interested in the mathematics education of students. But there are significant differences in how the two communities, broadly speaking, see this enterprise. The mathematicians' interests, naturally enough, are directed primarily at the graduate and undergraduate levels, and perhaps somewhat at the secondary level. In contrast, the interests of the educators are predominantly at the primary through secondary levels. But there is a broadening overlap in the ranges of

Mathematics and Teaching

interest of the two communities. And some fundamental aspects of what constitutes quality teaching are arguably independent of educational levels. Nonetheless, in areas of common focus, there are often profound differences of perspective and understanding across the two communities about what constitutes quality mathematics instruction. I expand on this below.

There is also a difference in educational priorities between the mathematics and mathematics education communities, a difference whose importance cannot be overstated. Mathematicians are naturally interested in "pipeline" issues, the rejuvenation of the professional community, and so the induction and nurturing of talented and highly motivated students into high level mathematical study. Mathematics educators, on the other hand, are professionally committed to the improvement of public education at scale, with the aim of *high levels of learning for all students*, and with a heavy emphasis on the word "all." While these two agendas are not intrinsically in conflict or competition, they are often seen to be so, and this can have resource and policy implications. Public education in the U.S., despite long and costly interventions, continues to perform poorly, in international comparisons, in terms of meeting workplace demands, and even in providing basic literacies. Moreover, there is a persistent achievement gap reflected in underperformance of certain subpopulations that the educational system historically has not served well. These are the "big frontline problems" of mathematics education, and they are just as compelling and urgent to mathematics educators as the big research problems are to mathematicians.

Mathematicians have an excellent tradition of nurturing students of talent. What they are less good at is identifying potential talent. The usual indices, high test scores and precocious accomplishment, are easy enough to apply, but these will typically overlook students of mathematical promise whom the system has not encouraged, or given either the expectation of or opportunity for high performance. As a result, mathematical enrichment programs, if not sensitively designed, can sometimes perpetuate the very inequities that mathematics educators are trying to mitigate.

My main focus here, however, is on teaching. Much of what I have learned I owe to work with my colleague and collaborator, Deborah Ball. To facilitate what I want to say, it will be helpful to use a schematic developed by Deborah and her colleagues David K. Cohen and Stephen W. Raudenbush (2003) to describe the nature of instruction.

The "Instructional Triangle"

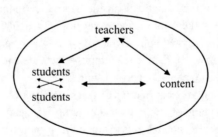

The concept proposed by this image is that instruction is about the interactions of the teacher with students (and of students with each other) around content. And the double arrows emphasize that these interactions are dynamic; in particular, changing any one element of the picture significantly alters the whole picture.

Now what I suggest is that most mathematicians' conception of instruction lives primarily at the content corner of the triangle, with its school incarnation expressed in terms of curriculum (including both standards documents and curriculum materials). In this view, with high quality curriculum in place, the mathematically competent teacher has only to implement that curriculum with intellectual fidelity, and then attentive and motivated students will learn. Of course mathematicians have little direct influence on schoolchildren, but they often have explicit, discipline-inspired ideas about what mathematics children should learn, and how. In the case of teachers, mathematicians do bear some direct responsibility for their mathematical proficiency, since many teachers learn much of the mathematics they know in university mathematics courses. Many studies have pointed to weak teacher content knowledge as a major source of underperformance in public education. Of the many remedies proposed, mathematicians have generally favored more and higher level mathematics courses as a requirement for certification. Unfortunately, none of the interventions based on these ideas so far undertaken have yet produced the desired gains in student achievement for a broad range of students. Note that all of this discussion and debate resides on the content-teacher side of the triangle, with little explicit attention yet given to the role of the students in instruction. Effective teachers both teach math and teach children.

An educational counterpart to the above "one-sided" (teacher-content) stance is an intellectually adventurous, but somewhat romantic view of school teaching, this time on the teacher-student side of the triangle. This approach proposes offering tangible and somehow "real-world" related mathematical activities with which to engage learners, but leaving the development of mathematical ideas to largely unrestrained student imagination and invention. In such cases the discipline of the mathematical ideas may be softened to the point of dissipation.

The most refined understanding of mathematics teaching, of which Deborah Ball's work is exemplary, insists that teaching must coordinate attention to the integrity of the mathematics, and appropriate learning goals, with attention to student thinking, which needs to be honored and made an integral part of the instruction. In this view, the effective teacher has not only a deep understanding of and fascination with the mathematics being taught, but also a dual knowledge of, and fascination with, student thinking. The underlying premise is that children have significant mathematical ideas, albeit imperfectly expressed, and that a part of the teacher's work is to recognize the presence of those ideas (for which a sophisticated knowledge of mathematics is needed!), to give them appropriate validation, and to help students shape them into a more developed articulation and understanding. It is the coordination of these dual spheres of attention—both to the mathematics and to students' thinking—that makes effective teaching the intricate and skilled work that it is.

This is a kind of professional practice that discursive rhetoric cannot adequately capture. It is best conveyed through an examination of teaching practice itself. So let me offer a vignette. Consider the teaching of mathematical proving. This is typically first done in high school geometry, but there are good reasons to argue that it should be done developmentally, starting in the early grades. What might this mean, or look like? After all, young children have no concept of anything approaching mathematical proof, and no one seriously argues that this should be formally taught to them. First of all, what is the intellectual imperative for proving that can be made meaningful to young children? Our view is that it is the persistent question, "Why are things true?" Deductive proof is the refined method devised by mathematicians to answer that very question. Learning the methods of deductive proving takes time, but the question—"Why are things true?"—is itself immediately compelling. The underlying pedagogy is that students progressively learn the methods of constructing compelling mathematical arguments in the course of repeatedly trying to convince others of things they have good reasons to believe to be true. In other words, students can be helped to construct the infrastructure of proving in the very course of proving things (to the best of their growing abilities).

To illustrate this idea, I turn to an episode from a third grade class (8-year-olds). The children have been exploring even and odd numbers. Although they do not yet have formal definitions, they rely on intuitive ideas of even and odd based on notions of fair-sharing. And they can identify whether particular (small) numbers are even or odd. They begin to notice addition patterns, like: even + even = even; even + odd = odd; and odd + odd = even. The teacher asks them whether they can "prove" that "Betsy's Conjecture" (odd + odd = even) is *always* true. Some of the children have tested the conjecture with lots of numbers, and they are thereby convinced that it must be true. The teacher challenges them, "How do you know that someone might not come up with a new example that didn't work?" The children are left to ponder that, some of them working collaboratively in small groups.

The next day, the teacher asks the class what they found out. Several hands went up, and she called on Jeanie, who, speaking also for her partner, Sheena, said, "*We were trying to prove that ... you can't prove that Betsy's Conjecture always works. ... Because numbers go on and on forever, and that means odd numbers and even numbers go on forever and, um, so you couldn't prove that all of them work.*" This stunning assertion is at once insightful and revolutionary. Jeannie's comment reflects the fact that Betsy's Conjecture is not just one, but rather infinitely many claims, and so not susceptible to empirical verification. Though articulated somewhat informally, she has realized that the cases of the conjecture go on and on.

A second student, Mei, objects to this argument. She objects not to the reasoning, but rather to an important inconsistency. She points out that prior claims for which the class had achieved consensus were also not checked in all cases, and Jeannie and her partner had not raised any similar objection to those earlier claims. She challenges the need to make sure that the claim works for all cases, since they had not tried to do this for other conjectures.

In both of these contributions, one can see the process by which the class is beginning to construct the very norms of mathematical reasoning, rejecting empirical methods for an infinitely quantified claim in the first instance, and the need for consistency of logical methods in the second. And all of this was still short of proving Betsy's Conjecture.

But the next day, Betsy produced a "proof" of her conjecture. At the board, she drew a row of seven small circles, and then to the right another row of seven small circles. Then she partitioned the two rows into three groups of two, with one left over, and the ones left over next to each other in the middle.

$$\underline{o\ o}\ \underline{o\ o}\ \underline{o\ o}\ \underline{o}\quad \underline{o}\ \underline{o\ o}\ \underline{o\ o}\ \underline{o\ o}$$

Then she turned to the class, and explained haltingly, "If you have an odd number and group it by twos you have one left over, so if you have two odd numbers and group them by twos, then the ones left over can be grouped together so you have an even number." What is most significant about this oral argument is she made no reference to the specific numbers, 7 and 7, that she had drawn, but rather spoke generically of "an odd number."

How did this demonstration surmount the dilemma raised by Jeannie and Sheena? Betsy had implicitly invoked a general definition of odd number—a number that when grouped by twos has one left over—and this definition, being itself infinitely quantified (it characterizes *all* odd numbers), can support a conclusion of similarly infinite purview. Betsy's oral argument might be algebraically expressed in the form

$$(x2 + 1) + (1 + y2) = x2 + (1 + 1) + y2 = (x + 1 + y)2$$

(Note that, for children at this stage, $2x = x + x$ is not the same as $x2 = 2 + 2 + \cdots + 2$.) But, although she was thinking in general terms, Betsy did not yet know how to represent her ideas generally using algebraic notation.

Notice for each of these students—Jeannie, Mei, and Betsy—the significance of the mathematical ideas, albeit sometimes expressed with difficulty. Episodes like this demonstrate what it might look like to develop the norms of proving in the course of constructing more and more well-developed arguments to convince others of the truth of claims. But this is not the whole story. This narrative resides, as does much of the education research literature, on the student-content side of the instructional triangle, focusing on accomplished student performance. The teacher seems invisible.

The above kind of collective interaction and learning does not occur spontaneously, unmediated by well-informed and purposeful instruction. The question of what makes for effective teaching has to do with understanding the knowledge, skills, and sensibilities (both mathematical and pedagogical), and above all, the practices, that enable instruction that can elicit the kind of motivation and reflective engagement of students described in the episode above. The kind of mathematical knowledge required includes not only a robust understanding of the mathematical terrain of the work, and corresponding learning goals, but also an ability to hear, in incipient and undeveloped form significant, though often not entirely correct, mathematical ideas in student thinking. I emphasize that the latter entails a special kind of knowledge of mathematics, not just psychology. Teaching requires the skills to not only hear and validate, and give space to these ideas, but also to help students reshape them in mathematically productive ways. The teacher needs to know how to instruct with questions more than with answers, and to give students appropriate time, space, and resources to engage these questions. The actions to accomplish this, though purposeful, structured, and carefully calibrated to the students, are also subtle, deliberately focusing on student performance. And so, to the naive observer, it often can appear that "the teacher is not doing much." The central problem of teacher education is to provide teachers with the knowledge and skills of such effective practice.

The idea of developmental learning, such as we saw with the third graders above, was also a feature of my calculus course back at Princeton in 1951. I first learned in that course what (a strange thing) a real number was (mathematically),[1] even though I had been working comfortably with real numbers throughout the upper high school grades. With regard to proving, while we learned, through witnessing stunning examples and through

---

[1] An equivalence class of Cauchy sequences of rational numbers.

practice problems, how to construct reasonably rigorous proofs, it was still always the case that the claims in question seemed at least reasonable, and sufficiently meaningful that general mathematical intuition could guide us. But I was at first stymied by a problem on one of our take-home exams. A function $f(x)$ was defined by: $f(x) = 0$ for x irrational, and $f(x) = 1/q$ if x is rational, equal to p/q in reduced form. The question was, "Where is f continuous?" At first sight this question seemed outrageous. How could one possibly answer it? It was impossible to sketch the graph. Intuition was useless. After some reflection I resigned myself to the fact that all we had to work with were definitions, of f, and of continuity. It was a great revelation to me that these definitions, and a modest amount of numerical intuition, sufficed to answer the question. This was a great lesson about the nature, and the power, of deductive reasoning, to which my enculturation, though at a different level, was not so different from that of the third graders above.

The construction and timing of such a problem was itself a piece of instruction. I wish now that we had video records of Emil Artin's calculus teaching that I could show to an accomplished teacher, like Deborah, to analyze the pedagogical moves, including interaction with students, of which his practice was composed. Much as I profited from that instruction, I was not prepared to see the craft of its construction.

References

Ball, D. L., & Bass, H. (2000). Making believe: The collective construction of public mathematical knowledge in the elementary classroom. In D. Philips (Ed.), *Constructivism in education: Yearbook of the National Society for the Study of Education* (pp. 193–224). Chicago: University of Chicago Press.

Ball, D. L., Hill, H.C., & Bass, H. (2005). Knowing mathematics for teaching: Who knows mathematics well enough to teach third grade, and how can we decide? *American Educator*, 29(1), p. 14–17, 20–22, pp. 43–46.

Ball, D. L., Thames, M.H., & Phelps, G. (2008). Content knowledge for teaching: What makes it special? *Journal of Teacher Education,* 59(5), pp. 389–407.

Cohen, D. K., Raudenbush, S. W., and Ball, D. L. (2003), Resources, Instruction, and Research, Educational Evaluation and Policy Analysis, 25(2), pp. 119–142.

Lakatos, I. (1976). *Proofs and refutations: The logic of mathematical discovery.* Cambridge University Press.

Thurston, William P. (1994). On proof and progress in mathematics, *Bulletin of the American Mathematical Society* 30(2): 161–177.

Varadarajan, V. S. (2010). I am a mathematician, (essay in this collection).

Weil, André (1971). The future of mathematics, in *Great Currents of Mathematical Thought*, Ed. F. Le Lionnais, vol. 1, Dover Publ., New York, pp. 321–336.

Wigner, Eugene (1960). The unreasonable effectiveness of mathematics in the natural sciences. *Communications in Pure and Applied Mathematics*, 13(1), pp. 1–14.

12

Who We Are and How We Got That Way?

Jonathan M. Borwein*

Abstract The typical research mathematician's view of the external world's view of mathematicians is more pessimistic and less nuanced than any objective measure would support. I shall explore some of the reasons why I think this is so. I submit that mathematics is a "science of the artificial" [18] and that we should wholeheartedly embrace such a positioning of our subject.

1 Putting Things in Perspective

> All professions look bad in the movies... why should scientists expect to be treated differently?
>
> —Michael Crichton[1]

I greatly enjoyed Steve Krantz's article in this collection that he showed me when I asked him to elaborate what he had in mind. I guess I am less pessimistic than he is. This may well reflect the different milieus we have occupied. I see the same glass but it is half full.

Some years ago, my brother Peter surveyed other academic disciplines. He discovered that students who complain mightily about calculus professors still prefer the relative certainty of what we teach and assess to the subjectivity of a creative writing course or the rigors of a physics or chemistry laboratory course. Similarly, while I have met my share of micro-managing Deans—who view mathematics with disdain when they look at the size of our research grants or the infrequency of our patents—I have encountered more obstacles to mathematical innovation within than without the discipline.

I do wish to aim my scattered reflections in generally the right direction: I am more interested in issues of creativity à la Hadamard [4] than in Russell and foundations or Piaget

---

* Centre for Computer Assisted Mathematics and its Applications (CARMA), School of Mathematical and Physical Sciences, University of Newcastle, NSW, Australia Email: jborwein@newcastle.edu.au Research supported by the Australian Research Council.

[1] Addressing the 1999 AAAS Meetings, as quoted in *Science* of Feb. 19, 1999, p. 1111.

and epistemology... and I should like a dash of "goodwill computing" thrown in. More seriously, I wish to muse about how we work, what keeps us going, how the mathematics profession has changed, how "*la plus ça change, la plus ça reste la même*,"[2] and the like while juxtaposing how we perceive these matters and how we are perceived. Elsewhere, I have discussed at length my own views about the nature of mathematics from both an aesthetic and a philosophical perspective (see, e.g., [10, 19]). I have described myself as "a computer-assisted quasi-empiricist." For present more psychological proposes I will quote approvingly from [5, p. 239]:

> ... Like Ol' Man River, mathematics just keeps rolling along and produces at an accelerating rate "*200,000 mathematical theorems of the traditional handcrafted variety... annually.*" Although sometimes proofs can be mistaken—sometimes spectacularly—and it is a matter of contention as to what exactly a "proof" is—there is absolutely no doubt that the bulk of this output is correct (though probably uninteresting) mathematics.
>
> —Richard C. Brown

Why do we produce so many unneeded results? In addition to the obvious pressure to publish and to have something to present at the next conference, I suspect Irving Biederman's observations below plays a significant role.

> "While you're trying to understand a difficult theorem, it's not fun," said Biederman, professor of neuroscience in the USC College of Letters, Arts and Sciences.... "But once you get it, you just feel fabulous."... The brain's craving for a fix motivates humans to maximize the rate at which they absorb knowledge, he said.... "I think we're exquisitely tuned to this as if we're junkies, second by second."
>
> —Irving Biederman[3]

Take away all success or any positive reinforcement and most mathematicians will happily replace research by adminstration, more and (hopefully better) teaching, or perhaps just a favorite hobby. But given just a little stroking by colleagues or referees and the occasional opiate jolt, and the river rolls on.

The pressure to publish is unlikely to abate and qualitative measurements of performance[4] are for the most part fairer than leaving everything to the whim of one's Head of Department. Thirty years ago my career review consisted of a two-line mimeo "*your salary for next year will be...*" with the relevant number written in by hand. At the same time, it is a great shame that mathematicians have a hard time finding funds to go to conferences just to listen and interact. Csikszentmihalyi [6] writes:

> [C]reativity results from the interaction of a system composed of three elements: a culture that contains symbolic rules, a person who brings novelty into the symbolic

---

[2] For an excellent account of the triumphs and vicissitudes of Oxford mathematics over eight centuries see [8]. The description of Haley's ease in acquiring equipment (telescopes) and how he dealt with inadequate money for personnel is by itself worth the price of the book.

[3] Discussing his article in the *American Scientist* at www.physorg.com/news70030587.html.

[4] For an incisive analysis of citation metrics in mathematics I thoroughly recommend the recent IMU report and responses at: openaccess.eprints.org/index.php?/archives/417-Citation-Statistics-International-Mathematical-Union-Report.html.

domain, and a field of experts who recognize and validate the innovation. All three are necessary for a creative idea, product, or discovery to take place.
—Mihalyy Csikszentmihalyi

We have not paid enough attention to what creativity is and how it is nurtured. Conferences need audiences and researchers need feedback other than the mandatory *"nice talk"* at the end of a special session. We have all heard distinguished colleagues mutter a stream of criticism during a plenary lecture only to proffer *"I really enjoyed that"* as they pass the lecturer on the way out. A communal view of creativity requires more of the audience.

2 Who We Are

As to who we are? Sometimes we sit firmly and comfortably in the sciences. Sometimes we practice—as the *Economist* noted—the most inaccessible of the arts[5] possessed in Russell's terms [17, p. 60] of *"a supreme beauty—a beauty cold and austere."* And sometimes we sit or feel we sit entirely alone. So forgive me if my categorizations slip and slide a bit. Even when we wish to remove ourselves from the sciences—by dint perhaps of our firm deductive underpinnings—they are often more than welcoming. They largely fail to see the stark deductive/inductive and realist/idealist distinctions that reached their apogee in the past century.

Yet many scientists have strong mathematical backgrounds. A few years ago I had the opportunity to participate as one of a team of seven scientists and one humanist who were mandated to write a national report on Canada's future need for advanced computing [14]. Five of us had at least an honors degree in mathematics. At the time none of us (myself included) lived in a mathematics department. The human genome project, the burgeoning development of financial mathematics, finite element modeling, Google and much else have secured the role of mathematics within modern science and technology research and development as *"the language of high technology"*; the most sophisticated language humanity has ever developed. Indeed, in part this scientific ecumenism reflects what one of my colleagues has called *"an astonishing lack of appreciation for how mathematics is done."* He went on to remark that in this matter we are closer to the fine arts.

Whenever I have worked on major interdisciplinary committees, my strong sense has been of the substantial respect and slight sense of intimidation that most other quantitative scientists have for mathematics. I was sitting on a multi-science national panel when Wiles's proof of Fermat's last theorem was announced. My confreres wanted to know *"What, why and how?"* "What" was easy, as always "why" less so, and I did not attempt "how." In [10] I wrote

> While we mathematicians have often separated ourselves from the sciences, they have tended to be more ecumenical. For example, a recent review of *Models. The Third Dimension of Science*[6] chose a mathematical plaster model of a Clebsch

---

[5] In "Proof and Beauty," *Economist* article, 31 Mar 2005. *"Why should the non-mathematician care about things of this nature? The foremost reason is that mathematics is beautiful, even if it is, sadly, more inaccessible than other forms of art."*

[6] See Julie K. Brown, Solid tools for visualizing science, *Science,* November 19, 2004, 1136–37.

diagonal surface as its only illustration. Similarly, authors seeking examples of the aesthetic in science often choose iconic mathematics formulae such as $e = mc^2$.

"How" is not easy even within mathematics. *A Passion for Science* [21] is the written record of thirteen fascinating BBC interviews with scientists including Nobelist Abdus Salam, Stephen Jay Gould, Michael Berry, and Christopher Zeeman. The communalities of their scientific experiences far outstrip the differences. Zeeman tells a nice story of how his Centre's administrator (a non mathematician) in Warwick could tell whether the upcoming summer was dedicated to geometry and topology, to algebra, or to analysis— purely on the basis of their domestic arrangements and logistics. For instance algebraists were very precise in their travel plans, topologists very inclusive in their social group activities and analysts were predictably unpredictable. I won't spoil the anecdote entirely but it reinforces my sense that the cognitive differences between those three main divisions of pure mathematics are at least as great as those with many cognate fields. In this taxonomy I am definitely an analyst—not a geometer or an algebraist.

There do appear to be some cognitive communalities across mathematics. In [4] my brother with Peter Liljedahl and Helen Zhai reported on the responses to an updated version of Hadamard's questionnaire [13] that they circulated to a cross-section of leading living mathematicians. This was clearly a subject the target group wanted to speak about. The response rate was excellent (over 50%) and the answers striking. According to the survey responses, the respondents placed a high premium on serendipity—but as Pasteur observed *"fate favours the prepared mind."* Judging by where they said they have their best ideas they take frequent showers and like to walk while thinking. They don't read much mathematics, preferring to have mathematics explained to them in person. They much more resemble theorists throughout the sciences than careful methodical scholars in the humanities.

My academic life started in the short but wonderful infusion of resources for science and mathematics "after sputnik"—I started University in 1967—and now includes the *Kindle Reader* (on which I am listening[7] to a fascinating new biography of the *Defense Advanced Research Projects Agency*, DARPA). The tyranny of a Bourbaki-dominated curriculum has been largely replaced by the scary gray-literature world of *Wikipedia* and *Google scholar*.

While typing this paragraph I went out on the web and found the Irving Layton poem, that I quote at the start of Section 4, in entirety within seconds (I merely googled "*And me happiest when I compose poems*"—I know the poem is *somewhere* in my personal library). For the most part this has been a wonderful journey. Not everything has improved from that halcyon pre-post-structuralist period a half-century ago when algebraists could command more attention from funding agencies than could engineers as [5] recalls. But the sense of time for introspection before answering a colleague's wafer-thin "airmail letter" enquiry, and the smell of mold that accompanied leisurely rummaging in a great library's stacks are losses in my personal life that measure the receding role of the University as the "*last successful medieval institution.*" [11]

---

[7] It will read to you in a friendly if unnatural voice.

2.1 Stereotypes from without Looking in

One of the epochal events of my childhood as a faculty brat in St. Andrews, Scotland was when C. P. Snow (1905–1980) delivered an immediately controversial 1959 Rede Lecture in Cambridge entitled "The Two Cultures."[8] Snow argued that the breakdown of communication between the "two cultures" of modern society—the sciences and the humanities—was a major obstacle to solving the world's problems—and he had never heard of global warming. In particular, he noted the quality of education was everywhere on the decline. Instancing that many scientists had never read Dickens, while those in the humanities were equally non-conversant with science, he wrote:

> A good many times I have been present at gatherings of people who, by the standards of the traditional culture, are thought highly educated and who have with considerable gusto been expressing their incredulity at the illiteracy of scientists. Once or twice I have been provoked and have asked the company how many of them could describe the Second Law of Thermodynamics, the law of entropy. The response was cold: it was also negative. Yet I was asking something which is about the scientific equivalent of: *'Have you read a work of Shakespeare's?'*

The British musical satirists Michael Flanders and Donald Swann took immediate heed of this for their terrific monologue and song "First and Second Law of Thermodynamics" that I can still recite from memory.

> [Michael:] Snow says that nobody can consider themselves educated who doesn't know at least the basic language of Science. I mean, things like Sir Edward Boyle's Law, for example: the greater the external pressure, the greater the volume of hot air. Or the Second Law of Thermodynamics—this is very important. I was somewhat shocked the other day to discover that my partner not only doesn't know the Second Law, he doesn't even know the First Law of Thermodynamics.
>
> Going back to first principles, very briefly, thermodynamics is of course derived from two Greek words: *thermos*, meaning hot, if you don't drop it, and *dinamiks*, meaning dynamic, work; and thermodynamics is simply the science of heat and work and the relationships between the two, as laid down in the Laws of Thermodynamics, which may be expressed in the following simple terms...
>
> After me...
>
> The First Law of Thermodymamics:
> Heat is work and work is heat
> Heat is work and work is heat
> Very good!
> The Second Law of Thermodymamics:
> Heat cannot of itself pass from one body to a hotter body
>   (scat music starts)
> Heat cannot of itself pass from one body to a hotter body

---

[8] Subsequently republished in [20].

> Heat won't pass from a cooler to a hotter
> Heat won't pass from a cooler to a hotter
> You can try it if you like but you far better notter
> You can try it if you like but you far better notter
> 'Cos the cold in the cooler will get hotter as a ruler
>
> ...

Snow goes on to say:

> I now believe that if I had asked an even simpler question—such as, *What do you mean by mass, or acceleration*, which is the scientific equivalent of saying, *"Can you read?"*—not more than one in ten of the highly educated would have felt that I was speaking the same language. So the great edifice of modern physics goes up, and the majority of the cleverest people in the western world have about as much insight into it as their Neolithic ancestors would have had.

C. P. Snow wrote pre-Kuhn, pre-Foucault, pre-much else [5]; and I submit that a half-century on the situation is worse, knowledge more fragmented, ignorance of science and mathematics more damaging to the public discourse.[9]

In addition, I think the problem was much less symmetric than Snow suggested. I doubt I have ever met a scientist who had not read (or at least watched on BBC) some Dickens, who never went to movies, art galleries, or the theatre. It is, however, ever more socially acceptable to be a scientific ignoramus or a mathematical dunce. It is largely allowed to boast "*I was never any good at mathematics at school.*" I was once told exactly that—in sotto voce—by the then Canadian Governor General during a formal ceremony at his official residence in Ottawa. Even here we should be heedful not to over-analyse as we are prone to do. Afterwards the "GG" (as Canadians call their Queen's designate) ruminated apologetically that if he had been a bit better at mathematics he would not have had to become a journalist. Some of this has been "legitimated" by denigrating science as "reductionist" and incapable of the deeper verities [5].

As Underwood Dudley has commented, no one apologizes for not being good at geology in school. Most folks understand that failing "Introduction to Rocks" in Grade Nine does not knock you off of a good career path. The outside world knows several truths: mathematics is important, it is hard, it is usually poorly taught in school, and the average middle-class parent is ill-prepared to redress the matter. I have become quite hard-line about this. When a traveling companion on a plane starts telling me that "*Mathematics was my worst subject in school.*" I will reply "*And if you were illiterate would you tell me?*" They usually take the riposte fairly gracefully.

Consider two currently popular TV dramas *Numb3rs* (mathematical) and *House* (medical). A few years ago a then colleague, a distinguished pediatrician, asked me whether I

---

[9] I can't resist including the following email anecdote:

> This morning Al Gore gave the "keynote" speech at SC09 (the largest annual supercomputing meeting). During the question-answer period, he mentioned a famous talk "The Two Cultures" about the lack of communication between science and humanities, by one Chester (??)—he drew a blank as to who it was. Sitting on the third row, I shouted out "Snow" (meaning C. P. Snow). One other person also shouted "Snow," and so Gore acknowledged that it was indeed Snow.

watched Num3rs. I replied *"Do you watch House? Does it sometimes make you cringe?"* He admitted that it did but he still watched it. I said the same was true for me with Numb3rs, that my wife loved it and that I liked lots about it. It made mathematics seem important and was rarely completely off base. The lead character, Charlie, was brilliant and good-looking with a cute smart girl friend. The resident space-cadet on the show was a physicist not a mathematician. What more could one ask for? Sadly for many of our colleagues the answer is *"absolute fidelity to mathematical truth in every jot and tittle."* No wonder so many of us make a dog's-breakfast of the opportunities given to publicize our work!

> Caution, skepticism, scorn, distrust and entitlement seem to be intrinsic to many of us because of our training as scientists.
>
> —Stephen Rosen[10]

To *"Caution, skepticism, scorn, distrust and entitlement,"* I'd add *"persistence, intensity, a touch of paranoia, and a certain lack of sartorial elegance"* but I still would not have identified mathematicians within the larger scientific herd. I think we are more inward drawn than theorists in, say, biology or physics. Our terminologies are more speciated between subfields and so we typically graze in smaller groups. But we are still bona fide scientists—contrary to the views of some laboratory scientists and some of our own colleagues.

> This is the essence of science. Even though I do not understand quantum mechanics or the nerve cell membrane, I trust those who do. Most scientists are quite ignorant about most sciences but all use a shared grammar that allows them to recognize their craft when they see it. The motto of the Royal Society of London is "Nullius in verba": trust not in words. Observation and experiment are what count, not opinion and introspection. Few working scientists have much respect for those who try to interpret nature in metaphysical terms. For most wearers of white coats, philosophy is to science as pornography is to sex: it is cheaper, easier, and some people seem, bafflingly, to prefer it. Outside of psychology it plays almost no part in the functions of the research machine.
>
> —Steve Jones[11]

2.2 Stereotypes from within Looking Out

Philosophy (not to mention introspection) is arguably more important to, though little more respected by, working mathematicians than it is to experimental scientists.

> Whether we scientists are inspired, bored, or infuriated by philosophy, all our theorizing and experimentation depends on particular philosophical background assumptions. This hidden influence is an acute embarrassment to many researchers, and it is therefore not often acknowledged. Such fundamental notions as reality, space,

---

[10] An astrophysicist, turned director of the *Scientific Career Transitions Program* in New York City, giving job-hunting advice in an on-line career counseling session as quoted in *Science*, August 4 1995, p. 637. He continues that these traits hinder career change!

[11] From his review of *How the Mind Works* by Steve Pinker, in the *New York Review of Books*, pp. 13–14, Nov 6, 1997.

time, and causality–notions found at the core of the scientific enterprise—all rely on particular metaphysical assumptions about the world.

—Christof Koch[12]

As I alluded to above, working mathematicians—by which I mean those of my personal or professional acquaintance—are overinclined by temperament and training to see meaning where none is intended and patterns where none exist. For the most part over the past centuries this somewhat autistic tendency has been a positive adaptation. It has allowed the discipline to develop the most powerful tools and most sophisticated descriptive language possessed by mankind. But as the nature of mathematics changes we should be heedful of Napoleon's adage *"Never ascribe to malice that which is adequately explained by incompetence,"*[13] or as Goethe (1749–1832) put it in [12]:

> Misunderstandings and neglect occasion more mischief in the world than even malice and wickedness. At all events, the two latter are of less frequent occurrence.

Suppose for "malice/wickedness" we substitute "meaning/reason" and likewise replace "neglect/misunderstandings" by "chance/randomness." Then these squibs provide an important caution against seeing mathematical patterns where none exist. They offer equally good advice when dealing with Deans.

3 Changing Modes of Doing Mathematics

Goethe's advice is especially timely as we enter an era of intensive computer-assisted mathematical data-mining; an era in which we will more and more encounter unprovable truths and salacious falsehoods. In [10] I wrote

> It is certainly rarer to find a mathematician under thirty who is unfamiliar with at least one of `Maple`, `Mathematica` or `MatLab`, than it is to find one over sixty five who is really fluent. As such fluency becomes ubiquitous, I expect a re-balancing of our community's valuing of deductive proof over inductive knowledge.

As we again become comfortable with mathematical *discovery* in Giaquinto's sense of being *"independent, reliable and rational"* [9], assisted by computers, the community sense of a mathematician as a producer of theorems will probably diminish to be replaced by a richer community sense of mathematical understanding. It has been said that Riemann proved very few theorems and even fewer correctly and yet he is inarguably one of the most important mathematical, indeed scientific, thinkers of all time. Similarly most of us were warned off pictorial reasoning:

> A heavy warning used to be given [by lecturers] that pictures are not rigorous; this has never had its bluff called and has permanently frightened its victims into playing

---

[12] In "Thinking About the Conscious Mind," a review of John R. Searle's *Mind. A Brief Introduction*, Oxford University Press, 2004.

[13] I have collected variants old and new on the theme of over-ratiocination at www.carma.newcastle.edu/jb616/quotations.html.

for safety. Some pictures, of course, are not rigorous, but I should say most are (and I use them whenever possible myself).

—J. E. Littlewood [16, p. 53][14]

Let me indicate how much one can now do with good computer-generated pictures.

3.1 Discovery and Proof: Divide-and-Concur

In a wide variety of problems such as protein folding, 3SAT, spin glasses, giant Sodoku, etc., we wish to find a point in the intersection of two sets A and B where B is non-convex. The notion of "divide-and-concur" as described below often works spectacularly—much better than theory can currently explain. Let $P_A(x)$ and $R_A(x) := 2\, P_A(x) - x$ denote respectively the *projector* and *reflector* on a set A as illustrated in Figure 1. Then "divide-and-concur"[15] is the natural geometric iteration "reflect-reflect-average":

$$x_{n+1} \Longrightarrow \frac{x_n + R_A(R_B(x_n))}{2}. \tag{12.1}$$

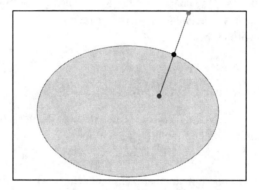

Figure 1. Reflector (interior) and Projector (boundary) of a point external to an ellipse.

Consider the simplest case of a line A of height α and the unit circle B [2]. With $z_n := (x_n, y_n)$ we have:

$$x_{n+1} := \cos(\theta_n),\ y_{n+1} := y_n + \alpha - \sin(\theta_n), \quad (\theta_n := \arg z_n). \tag{12.2}$$

This is intended to find a point on the intersection of the unit circle and the line of height α as shown in Figure 2 for $\alpha = .94$.

We have also studied the analogous differential equation since asymptotic techniques for such differential equations are better developed. We decided

$$x'(t) = \frac{x(t)}{r(t)} - x(t) \text{ where } r(t) := \sqrt{x(t)^2 + y(t)^2}\,] \tag{12.3}$$

$$y'(t) = \alpha - \frac{y(t)}{r(t)}$$

---

[14] Littlewood (1885–1977) published this in 1953, long before the current fine graphic, geometric, and other visualization tools were available.

[15] This is Cornell physicist Veit Elser's slick term for the algorithm in which the reflection can be performed on separate cpu's (divide) and then averaged (concur).

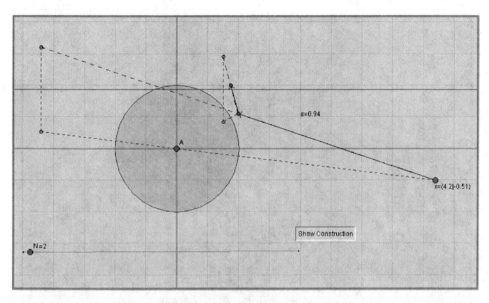

Figure 2. The first three iterates of (2) in *Cinderella*.

was a reasonable counterpart to the Cartesian formulation of (2)—we have replaced the difference $x_{n+1} - x_n$ by $x'(t)$, etc.—as shown in Figure 4.

Following Littlewood, I find it hard to persuade myself that the pictures in Figures 3 and 4 do not constitute a *generic proof* of the algorithms they display as implemented in an applet at users.cs.dal.ca/~jborwein/expansion.html. In Figure 3 we see the iterates spiralling in towards the right-hand point of intersection with those closest to the *y*-axis lagging behind but being unremittingly reeled in to the point. Brailey Sims and I have now found a conventional proof that the behavior is as observed [3] but we discovered all the results first graphically and were led to the appropriate proofs by the dynamic pictures we drew.

4 The Exceptionalism of Mathematics

> And me happiest when I compose poems.
> Love, power, the huzza of battle
> Are something, are much;
> yet a poem includes them like a pool.—Irving Layton [15, p. 189]

This is the first stanza of the Irving Layton (1912–2006) poem "The Birth of Tragedy." Explicitly named after Nietzsche's first book, Layton tussles with Apollonian and Dionysian impulses (reason versus emotion). He calls himself "A quiet madman, never far from tears" and ends "*while someone from afar off blows birthday candles for the world.*" Layton, who was far from a recluse, is one of my favourite Canadian poets.

I often think poetry is a far better sustained metaphor for mathematics than either music or the plastic arts. I do not see poetry making such a good marriage with any other science. Like good poets, good mathematicians are often slightly autistic observers of a somewhat

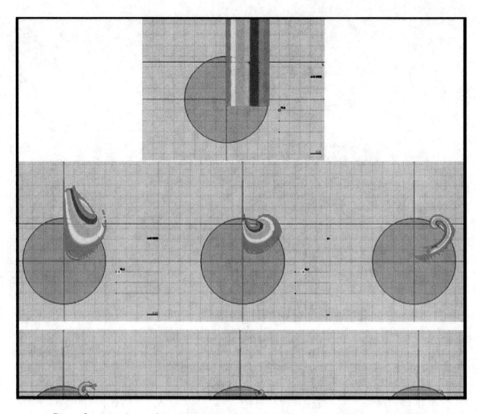

Figure 3. Snapshots of 10,000 points after 0, 2, 7, 13, 16, 21, and 27 steps of (2).

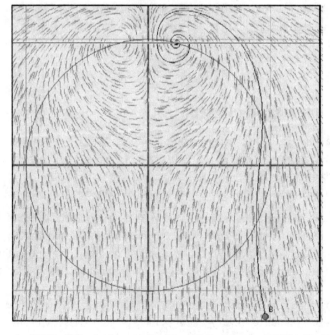

Figure 4. ODE solution and vector field for (12.3) with $\alpha = 0.97$ in *Cinderella*.

dysphoric universe. Both art forms at their best distill and concentrate beauty like no other and both rely on a delicate balance of form and content, semantics and syntax.

Like all academic disciplines we are (over-)sure of our own specialness.

- *Mathematicians are machines for turning coffee into theorems.* (Renyi)
- *A gregarious mathematician is one who looks at the other person's feet when addressing them.*
- *Mathematics is what mathematicians do late at night.*
- *You want proof. I'll give you proof.* (Harris)
- *There are three kinds of mathematician, those who can count and those who can't.*

Most of these can be—and many have been—used with a word changed here or there about statisticians, computer scientists, chemists, physicists, economists, and philosophers. For instance "*There are 10 kinds of computer scientists, those who understand binary and those who don't.*" It is amusing to ask colleagues in other sciences for their corresponding self-identifying traits. All of the above mentioned groups except the philosophers are pretty much reductionists:

> Harvard evolutionary psychologist Steven Pinker is probed on "*Evolutionary Psychology and the Blank Slate.*" The conversation moves from the structure of the brain to adaptive explanations for music, creationism, and beyond. Stangroom asks Pinker about the accusations that biological explanations of behavior are determinist and reduce human beings to the status of automatons.". . . "*Most people have no idea what they mean when they level the accusation of determinism,*" Pinker answers. "*It's a nonspecific 'boo' word, intended to make something seem bad without any content.*"[16]

Steve Jones is quoted in the same article equating philosophy and pornography and while many of us, myself included, see a current need to rethink the philosophy of mathematics, Pinker and he capture much of the zeitgeist of current science including mathematics.

4.1 Mathematics as a Science of the Artificial

Pure mathematics, theoretical computer science, and various cognate disciplines are sciences of the artificial in that they study *scientifically* man-made *artificial* concepts. Mathematical experiments and data collection are clearly not taking place in the natural world. They are at best quasi-empirical and yet they subscribe fully to the scientific method. Like other sciences they are increasingly engaged in "exploratory experimentation" [1, 2]. In *The Sciences of the Artificial* [18, p. 16] Herb Simon compellingly wrote about reductionism:

> This skyhook-skyscraper construction of science from the roof down to the yet unconstructed foundations was possible because the behavior of the system at each level depended only on a very approximate, simplified, abstracted characterization

---

[16] *The Scientist* of June 20, 2005 describing Jeremy Stangroom's interviews in *What (some) scientists say*, Routledge Press, 2005.

at the level beneath. This is lucky, else the safety of bridges and airplanes might depend on the correctness of the "Eightfold Way" of looking at elementary particles.

> More than fifty years ago Bertrand Russell made the same point about the architecture of mathematics. See the Preface to *Principia Mathematica* "... *the chief reason in favor of any theory on the principles of mathematics must always be inductive, i.e., it must lie in the fact that the theory in question allows us to deduce ordinary mathematics. In mathematics, the greatest degree of self-evidence is usually not to be found quite at the beginning, but at some later point; hence the early deductions, until they reach this point, give reason rather for believing the premises because true consequences follow from them, than for believing the consequences because they follow from the premises.*

Contemporary preferences for deductive formalisms frequently blind us to this important fact, which is no less true today than it was in 1910.

I love the fact that Russell the arch-deductivist so clearly describes the fundamental role of inductive reasoning within mathematics. This long-but-rewarding quote leads me to reflect that we mathematicians need more strong-minded and assured critics. I acknowledge that it is easier to challenge a speaker in history or philosophy. One may reasonably disagree in a way that is hard in mathematics.[17] When someone stands up in a mathematics lecture and says she can answer the speaker's hard open question, nine times out of ten the respondent has misunderstood the question or misremembered her own prior work. We do, however, need to develop a culture that encourages spirited debate of such matters as how best to situate our subject within the academy, how important certain areas and approaches are, how to balance research and scholarship, and so on. Moreover, fear and lack of mutual respect for another's discipline make it hard to venture outside one's own niche. For instance, many physicists fear mathematicians who, in turn, are often most uncomfortable or dismissive of informal reasoning and of "physical or economic intuition."

4.2 Pure versus Applied Mathematics

Mathematics is at once both a set of indispensable tools and a self-motivating discipline; a mind-set and a way of thinking. In consequence there are many research mathematicians working outside mathematics departments and a smaller but still considerable number of non-mathematicians working within. What are the consequences? First, it is no longer possible to assume that all of one's colleagues could in principle—if not with enthusiasm or insight—teach all the mathematics courses in the first two years of the university syllabus. This pushes us in the direction of other disciplines like history or biology in which teaching has always been tightly coupled with core research competence.

At a more fundamental level, I see the discipline boundary as being best determined by answering the question as to whether the mathematics at issue is worth doing in its own

---

[17] Some years ago I persuaded Amazon to remove several unsubstantiated assertions about "errors on every page" in one of my books—by a digital groupie turned stalker—from their website after I pointed out that while one could have an opinion that a Cormac McCarthy novel was dull but assertions of factual error were subject to test.

right. If the answer is "yes," then it is "pure"[18] mathematics and belongs in the discipline; if not then, however useful or important the outcome, it does not fit. The latter would, for example, be the case of a lot of applied operations research, a good deal of numerical modeling and scientific computation, and most of statistics. All significant mathematics should be nourished within mathematics departments, but there are many important and useful applications that do not by that measure belong.

5 How to Become a Grownup Science

As Darwin [7] ruefully realized rather late in life, we mathematicians have a lot to offer:

> During the three years which I spent at Cambridge my time was wasted, as far as the academical studies were concerned, as completely as at Edinburgh and at school. I attempted mathematics, and even went during the summer of 1828 with a private tutor (a very dull man) to Barmouth, but I got on very slowly. The work was repugnant to me, chiefly from my not being able to see any meaning in the early steps in algebra. This impatience was very foolish, and in after years I have deeply regretted that I did not proceed far enough at least to understand something of the great leading principles of mathematics, for men thus endowed seem to have an extra sense.—Charles Darwin

We also have a lot to catch up with. We have too few accolades compared to other sciences: prize lectures, medals, fellowships and the like. We are insufficiently adept at boosting our own cases for tenure, for promotion, or for prizes. We are frequently too honest in reference letters. We are often disgracefully terse—unaware of the need to make obvious to others what is for us blindingly obvious. I have seen a Fields medalist recommend a talented colleague for promotion with the one line letter "*Anne has done some quite interesting work.*" Leaving aside the ambiguity of the use of the word "quite" when sent by a European currently based in the United States to a North American promotion committee, this summary is pretty lame when compared to a three page letter for an astrophysicist or chemist—that almost always tells you the candidate is the top whatever-it-is in the field. A little more immodesty in promoting our successes is in order.

I'm not encouraging dishonesty, but it is necessary to understand the ground rules of the enterprise and to make some attempt to adjust to them. When a good candidate for a Rhodes Scholarship turns up at ones office, it should be obvious that a pro forma scrawled note

> Johnny is really smart and got an "A+" in my advanced algebraic number theory class. You should give him a Rhodes scholarship.

is inadequate. Sadly, the only letters of that kind that I've seen in Rhodes scholarship dossiers have come from mathematicians.

I am a mathematician rather than a computational scientist or a computer scientist primarily because I savour the structures and curiosities (including *spandrels* and *exaptations*

---

[18] Which may well be highly applicable.

in Gould's words) of mathematics. I am never satisfied with my first proof of a result and until I have found limiting counterexamples and adequate corollaries will continue to worry at it. I like attractive generalizations on their own merits. Very often it is the unexpected and unintended consequences of a mathematical argument that when teased out provides the real breakthrough. Such often leads eventually to tangible and dramatic physical consequences: take quantum mechanical tunneling.

A few years ago I had finished a fine piece of work with a frequent collaborator who is a quantum field theorist—and a man of great insight and mathematical power. We had met success by introducing a sixth-root of unity into our considerations. I mooted looking at higher-order analogues. The reply came back *"God in her wisdom is happy to build the universe with sixth-roots. You, a mathematician, can look for generalizations if you wish."*

6 Conclusion

I became a mathematician largely because mathematics satisfied four criteria. (i) I found it reasonably easy; (ii) I liked understanding or working out how things function; but (iii) I was not much good with my hands and had limited physical intuition; (iv) I really disliked pipettes but I wanted to be a scientist. That left mathematics. Artificial yes, somewhat introspective yes, but informed by many disciplines and clearly an important science.

I have had several students whom I can not imagine following any other life path but I was not one of those. I would I imagine have been happily fulfilled in various careers of the mind; say as an historian or an academic lawyer. But I became a mathematician. It has been and continues to be a pretty wonderful life.

References

[1] Jonathan M. Borwein. "Exploratory Experimentation," column in "Maths Matters," *Australian Mathematical Society Gazette*, **36** (2009), 166–175.

[2] J. M. Borwein. "Exploratory Experimentation: Digitally-assisted Discovery and Proof." in *ICMI Study 19: On Proof and Proving in Mathematics Education*. Springer, 2012.

[3] Jonathan Borwein and Brailey Sims. "The Douglas-Rachford algorithm in the absence of convexity." Submitted, *Fixed-Point Algorithms for Inverse Problems in Science and Engineering*, to appear in *Springer Optimization and Its Applications*, November 2009.

[4] Peter B. Borwein and Peter Liljedahl and Helen Zhai. *Creativity and Mathematics*. MAA Spectrum Books, 2014.

[5] Richard C. Brown. *Are Science and Mathematics Socially Constructed?* World Scientific, 2009.

[6] Mihalyy Csikszentmihalyi, *Creativity: Flow and the Psychology of Discovery and Invention*. Harper Collins, 1997.

[7] Charles Darwin, *Autobiography of Charles Darwin*. Available at infomotions.com/etexts/gutenberg/dirs/etext99/adrwn10.htm.

[8] John Fauvel, Raymond Flood, and Robin Wilson (Eds.). *Oxford Figures. 800 Years of the Mathematical Sciences*. Oxford University Press, 1999.

[9] Marco Giaquinto. *Visual Thinking in Mathematics. An Epistemological Study*. Oxford Univ. Press, 2007.

[10] Bonnie Gold and Roger Simons. *Proof and Other Dilemmas.* MAA Spectrum Books, 2008.

[11] A. Bartlett Giametti, *A Free and Ordered Space: The Real World of the University.* Norton, 1990.

[12] Johann Wolfgang von Goethe, *The Sorrows of Young Werther.* Available in English at www.gutenberg.org/etext/2527.

[13] Jacques Hadamard, *The Mathematician's Mind: The Psychology of Invention in the Mathematical Field.* Notable Centenary Titles, Princeton University Press, 1996. First edition, 1945.

[14] Kerry Rowe et al., *Engines of Discovery: The 21st Century Revolution. The Long Range Plan for HPC in Canada.* NRC Press, 2005, revised 2007.

[15] by David Stouck, *Major Canadian Authors: A Critical Introduction to Canadian Literature in.* Nebraska University Press, 1988.

[16] J. E. Littlewood, *Littlewood's Miscellany*, Cambridge University Press, 1953. Revised edition, 1986.

[17] Bertrand Russell, "The Study of Mathematics" in *Mysticism and Logic: And Other Essays.* Longman, 1919.

[18] Herbert A. Simon. *The Sciences of the Artificial.* MIT Press, 1996.

[19] Nathalie Sinclair, David Pimm and William Higginson (eds.), *Mathematics and the Aesthetic: New Approaches to an Ancient Affinity.* CMS Books in Mathematics. Springer-Verlag, 2007.

[20] C. P. Snow, *The Two Cultures and the Scientific Revolution.* Cambridge University Press, 1993. First published 1959.

[21] Lewis Wolpert and Alison Richards (eds). *A Passion for Science.* Oxford University, Press, 1989.

13

Social Class and Mathematical Values in the USA

Roger Cooke

Abstract. The influence of socio-economic background on the choice of mathematics as a career and the intellectual values of the mathematician is examined through the lens of the author's life experience and an informal sampling of biographies of noteworthy mathematicians.

Introduction

Having read other articles in this collection, I have seen several analyses of the psychological traits that shape a mathematician. I would like to supplement that discussion with a discussion of the social and economic backgrounds of mathematicians. Unfortunately, I do not have the background in sociology, political science, or economics to do so. I am thereby forced to fall back on the one case I know best—myself. Perhaps my experience will reflect the experience of some others in the mathematical community. My aim is to make a connection between my socioeconomic background and my views of the role the mathematical community should play in American society. The reader is invited to imagine or write a similar article about himself or herself and will, I hope, forgive my talking about myself at such length.

My Socio-Economic Background

The reflections that follow are the observations of a mathematician who grew up and worked in the United States during the last half of the twentieth century. I do not claim that they are an accurate picture of the country as a whole during this period. I have not read any scholarly studies, much less done any of my own to support what I am about to say. To keep this article from being only the ruminations of a mathematician on the brink of senility, I performed a small-scale, highly non-scholarly study of the biographies of 85 mathematicians, looking for details of their family history that would give a clue to their socio-economic class. The results of that study will be discussed in the final section of this article.

Throughout what is now becoming a long life, I have noticed some differences in basic values among members of the mathematical community. I ascribe these differences to social class. My thesis is that many scholars—and most mathematicians, it seems to me—were raised with the expectation of having a career involving intellectual activity, in homes where abstract ideas were found in books that lined the walls and were discussed at family gatherings. I was not among those people. Our home at 2620 Salu Street in Alton, Illinois contained very few books, most of them comparable to Reader's Digest condensed editions. There were no works of scholarship that I can remember. Discussions with family and friends were always on very concrete current events, sports, and (in my adolescence) television programs.

What I am presenting here, as a tentative hypothesis that I think might be worth investigating if it has not already been researched, is a two-part proposition: (1) the likelihood that a person will eventually become a mathematician is strongly influenced by the intellectual environment of that person's childhood; (2) the social and ethical values instilled in that childhood will very likely continue to influence a person who does become a mathematician. Both of these claims seem plausible intuitively. I hope to illustrate them with some autobiography and a sampling of the biographies of noteworthy mathematicians. The sample that provides a test of the first of these claims forms the final section of this essay.

My Family

My parents were the first in their families to move off the farm and get work in the city. In 1940, my father paid $100 to learn how to weld, joined the pipefitters union at the Shell Oil Company refinery in nearby Roxana, and worked there until his retirement in 1969. My mother stayed home to raise her three sons.[1] I believe this migration was typical of the migration from the farms to the cities in the U.S.A. during the first half of the twentieth century, and my parents were probably very typical representatives of it. They were the third generation of their families to have lived in the United States and the second to have been born here. My father's ancestors came from the British Isles, my mother's from Germany and Switzerland.

My parents were still on their farms in the early 1930s, when the Great Depression began. Being one of the older children in a family with five daughters and four sons, my father had been shipped out to Kansas in 1929 to live with his aunt. His formal education ended at that point. He never saw the inside of a high school. And yet, as anyone who met him realized, he was exceptionally bright. He read voraciously and (as I found out only in my old age) loved poetry. He kept his more "refined" tastes concealed, I imagine, in order to fit in with the people he lived among and worked with. I have often wondered what he would have made of his life if he had been given the opportunities that were lavished upon

---

[1] I was the middle one, having been born on July 31, 1942. As a mathematician, I can't help noting that I was born 833 days after my older brother and 834 days before my younger brother. It appears my parents did believe in family planning.

me. He lived to be 92, and was still mentally sharp, playing golf and arguing politics—he was supporting Hillary Clinton's campaign at the time—two weeks before he died.

Although she finished high school, my mother never had any opportunity to exhibit whatever intellectual talent she possessed, being confined to a traditional homemaker's role. It was a role that she accepted, though she often confided to other women, within my hearing, that she found it a great burden. I cannot form any estimate of what she might have done with the opportunities young women now have. I imagine that her life differed comparatively little from the traditional Kinder-Küche-Kirche that must have been the lot of her female ancestors in Germany. I believe that this kind of gender role assignment, though it was nearly universal in past times, remained and remains more pronounced among those whom Marx would call the proletariat than among those he would call the bourgeoisie, and especially more so than among the intelligentsia.

Family Values, Midwestern-Style

My parents married in 1938 and received some economic support from their parents in the early years of their union. Having been shaped by their upbringing, education, and life experience, they both had very simple ideas of the good life. The good life required responsibility on the part of the individual: You got yourself a job, however humble, boring, and low-paying. You clung to that job, as to your most precious possession, lived on what you could make, and didn't go into debt for anything but a house. They were not risk-takers. Neither of them drank alcohol or smoked, perhaps from moral principle, but also from simple thrift. The thought of going into business for themselves never occurred to either of my parents, as far as I know.

Despite all that, my childhood was not the kind of life depicted in old movies about Welsh coal miners. My mother taught me to play the piano when I was 4 years old and really hoped I would become a professional musician. She even took in other people's laundry in order to pay for my private lessons with a good piano teacher. It was through that teacher that I became acquainted with the great composers of European music. So much sacrifice on my mother's part, and it went shamefully unappreciated by me! I resented the lessons and hated having to practice. Now, even as my fingers grow stiff with arthritis, I am terribly grateful that I can play the classic works that I know, and I practice willingly three times a week.

World War II and its Effects on American Life

By working in a vital defense industry and having three children, my father avoided the adventures that befell his younger brother Norman, who was aboard the battleship Utah when Pearl Harbor was attacked,[2] and his older brother Paul, who landed in France in 1944. My father was drafted and dutifully appeared for induction in mid-1945, only to learn that the war was essentially over and the draft had been suspended.

The end of the war meant a tremendous boom for the American economy. Savings that accumulated during the period of rationing sparked both inflation and industrial expansion. A Golden Age began for some Americans. Money was lavished on consumer goods, and

---

[2] He served the remainder of the war on the heavy cruiser Portland.

still there were funds left over to build roads, bridges, and schools. Being in school at the time, my contemporaries and I were the favored and spoiled children of a generation that had known real privation.

Class, Race, and Gender

I grew up in a family and community that were thoroughly racist. The schools in Alton, which was the home of the abolitionist martyr Elijah Lovejoy and the site of the seventh Lincoln–Douglas debate in 1858, were segregated until the end of my time in elementary school, when one token black student was admitted to my sixthgrade class.[3] We lived in a neighborhood that was a slim finger, two blocks wide and three blocks long, of white families interwoven with fingers of black families of comparable size to the north, south, and west. My brothers and I learned early on not to wander outside this white enclave. To do so meant threats and humiliation, and possibly physical violence. Though my parents would not admit to feeling any fear, somehow their uneasiness with the neighborhood communicated itself to the younger generation. Just after I graduated from high school, they joined the white flight and moved to a suburb, eventually settling on the farm that my grandfather inherited from his brother.[4]

My brothers and I and our friends saw only the white side of this situation and, as children, did not contemplate what it must have looked like from the other side. We never asked ourselves why we were so resented that children who did not even know us would bully and intimidate us. Nor did we ask why we walked past the Dunbar School every day, just two blocks from our house, to trek nearly a mile to the Horace Mann school. In 1950, black teachers and pupils struggling for integration came and sat in the halls for several weeks, conducting their classes and eating their lunches there. It was mostly a peaceful demonstration, but my second-grade teacher had a smile of contentment on her face the morning she announced to us that "they went back to their own schools."

If schools were the only site of racism in Alton, black children might not have been so disadvantaged economically. The real estate market discriminated against them, as white people struggled to maintain the purity of their neighborhoods. But the worst aspect of all was the discrimination in employment. My father, as I have said, belonged to the pipefitters union. That union was dedicated to protecting its members' jobs. To do so, it was willing to lock black people out.[5]

With their parents excluded from the more lucrative semi-skilled jobs, it is not surprising that fewer than half a dozen black pupils were in my college-preparatory course in chemistry.

---

[3] He lived in a house that was on my way home from school, and he and I often walked together and talked and enjoyed each other's company. During junior high school we were—without our knowledge or consent—"tracked" differently, and I lost touch with him until high school. There I was surprised to find that this boy with whom I had been friendly and who had been friendly with me, had turned into a hostile, angry adolescent. I now wonder why I found this surprising.

[4] They were no longer farmers. The land was farmed by a neighbor who paid rent in the form of a share of the proceeds from sale of the wheat, corn, and soybeans they raised.

[5] Although I deeply regret the destruction of labor unions that has taken place over the past 50 years, I cannot deny the sinister side these unions had. They were undemocratic and willing to use strong-arm tactics to keep their own members in line, and they were racist. The power of my father's union was broken by a long and unsuccessful strike in 1962.

The potential of many talented black students was blighted in a way that cannot now be corrected. At least one of my black classmates, however, has had both an honorable record of military service and a civilian career that I hope has brought him satisfaction. When I last saw him—now half a century ago—he had been briefly hired, under pressure from the black community, as a shoe salesman. He picked me up and gave me a ride downtown one day when I was home, and I learned that he had been quickly fired from that job. He eventually found himself in Viet Nam, after which he went to college and graduate school and launched a long and distinguished career in health care administration. But—as anyone will agree—the fact that remarkable individuals can overcome obstacles does not in any way reduce the urgent social need to remove those obstacles and to compensate for them to the extent possible.

Since I am discussing discrimination at this point, I cannot neglect the equally prevalent sexism in the time when I grew up. Although girls were not excluded from academic and professional careers, they were certainly discouraged from them by the social milieu of the times. Of my female classmates, I know only one, the late Julia Thompson (1942–2004), professor of physics at the University of Pittsburgh, who had a distinguished career in science. In the sample of 85 mathematicians that I discuss in the final section of this essay, randomizing resulted in the inclusion of eight women.[6]

Awareness of this aspect of my childhood, while it does not make me feel guilty for taking advantage of available opportunities, does often cause me to wonder: How well would I have done if I had had to compete with women and minorities on a level playing field?

My Mental Development

Besides my piano teacher, there were many other influences on me, mostly through the public school system in Alton, beckoning me out of the world that my parents had made. The resulting development of career goals meant a complete reorientation of my priorities and values. I grew away from my family. By the time I was in high school, I was eagerly reading the works of Thoreau, Emerson, and (most influential of all) Bertrand Russell. I was starting on that road to a better life that my parents intended for me, but it was not to be merely a fulfillment of what they wanted—better income and more interesting work. My ideals were changing as well. Suddenly, knowledge became a value for its own sake. I wanted to know things, everything. This kind of humanistic ideal, ars gratia artis, scientia gratia scientiæ, was very alien to my parents. My father, for example, once decried the study of history. The only value he could see in it was the one known from Santayana's aphorism, "Those who cannot remember the past are doomed to repeat it." And even that practical application of history, he thought, did not involve knowing anything that happened earlier than 1900.

---

[6] One of them, Etta Falconer (1933–2002), was an American of African descent. Her father was a physician. In this connection, I mention that the famous jazz musician Miles Davis was born in my home town—his family moved to East St. Louis when he was one year old. His father was a dentist.

Why Mathematics?

Given that my emotional center had moved from the milieu of family to the world of scholarship by the time I reached high school, one can still ask how it happened that I chose mathematics as a career. I am not a goaldirected person, and I tend to react to whatever happens rather than trying to achieve anything special. What happened that led me into mathematics?

I was fascinated mainly by two things when I was in high school: foreign languages and science. Two years of Latin gave me a modest ability to read Latin prose, and I was encouraged and thrilled by being able to read some things that earlier would have baffled me, and still baffled others. It was, to me, every boy's dream of knowing a secret code. I made some attempts (not very successful) to learn German and Russian on my own. When I was in college, I became obsessed with Russian, and the skills I developed there and have continued to hone over the past 50 years, have brought me both emotional and financial rewards.[7]

Even more than languages, I wanted to know science, especially chemistry and physics, and most especially the theory of relativity, which was, to Americans who lived in the 1950s, the very quintessence of an arcane scientific theory. I have never fulfilled this ambition to my own satisfaction. I thought I would work on it in retirement, but have made very little progress in the eight years since I retired, for reasons I will discuss below.

Why then, did I become a mathematician? I think the answer lies in remarks my father casually made without thinking of the effect they would have on me. Although he had a proletarian resentment of the young engineers hired at the Shell Oil refinery, with their "book-learning" and their ignorance of the actual functioning of the physical plant that was so familiar to him, he saw that their training gave them a potential that was closed to him. He often remarked that he wished he had learned the chemistry and physics that they knew. Especially, he said, one needed to know mathematics. That remark caused me to make learning mathematics a high priority, as background for the scientific career I was planning.

Mathematics also had an intrinsic fascination for me. In ninth grade, when I learned to graph linear equations, I was fascinated beyond words by the fact that two inconsistent equations would graph as parallel lines, having no point of intersection. The geometric property of parallelism corresponded to the algebraic property of inconsistency. I eagerly showed this fact to my father, who smiled weakly, not wishing to discourage me, and said that "it's no use trying to explain that to someone who never had anything but arithmetic." Later, when I encountered logarithms, a word that I knew from the reverential way a friend of the family used it, said friend having found them terribly difficult, I was fascinated once again by the fact that adding them corresponded to multiplying the numbers that generated them. Once again, the beauty of the analogy drew me in. I wanted to know these things—really understand them, not just be able to apply them haltingly and with great effort.

But I had other interests as well. The ideal held up for me by my teachers and the educational authorities of the time was that of the broadly educated citizen. One needed to know not only science, but also music and art and literature and history. All these areas

---

[7] Moonlighting as a translator of Russian scientific and mathematical works enabled me to finance my children's college educations. It was probably my linguistic skill as much as my scholarly record that eventually got me onto the Academy exchange with the USSR in 1988–89, making it possible for me to live in Moscow.

became fascinating to me, and I strove to become as broadly educated as possible. That striving has stayed with me during my entire life and prevented me from focusing on the kind of narrow, specific problems that a mathematician needs to study in order to have a successful career in research.[8] Even within mathematics, as every specialist knows, the prudent thing is to have a small circle of interesting problems to work on, not to master broad swaths of mathematical theory. I have always been constitutionally unable to be prudent in this way.

I still remember the mid-1950s when, delivering newspapers after school, I contemplated a future in which I would learn everything in the proper order: first logic, then mathematics, then physics, chemistry, and biology. I wanted to have an encyclopedia of science in my head organized according to this scheme. Of course, I now realize that human knowledge does not function this way. Each science has its own starting point(s), independent of the others, even when the different sciences interact fruitfully.

College and Graduate School

Social class played hardly any role among my classmates through high school. Although some of them came from families that were considerably wealthier than mine, there was no snobbery. I, at least, did not know anyone who went to a private boarding school. But this social fluidity did mask the fact that there were great differences in the way we were being prepared for college. Only about one-fourth of the students in my high school went on to a college or university. Most of that one-fourth had parents who had also been to college. Children raised by college graduates had some idea of what to expect at a university and would not themselves seem as strange to their college classmates as I did. A good example of this difference and the effect that social background can have is provided by the story of my college education. A family friend two years older than I had entered Northwestern University and was enthusiastic about it. His family encouraged me to apply there. I did, and was accepted, and eventually graduated from Northwestern. But there were some complications along the way, complications that reflect the social differences I am discussing.

My father saw college as something of a risk and urged me to play it safe. He thought I could attend Shurtleff College in Alton (now the Alton campus of Southern Illinois University), get a good background for two years while living at home, then apply to transfer to a better place. My teachers encouraged me to aim higher, being sure I would get a good scholarship (as in fact I did). One friend in the class ahead of me had gone to the University of Pennsylvania, as had his father. Hearing what I was planning, his father (an influential alumnus) immediately contacted Penn, and phoned me to say that he had arranged for a full-tuition scholarship for me at Penn, to which I had not even applied. All I had to do was say the word, and I would be in. I didn't know what to do.

At this point, my mother asserted herself in a way that she rarely did. Unable to bear the thought that I might move so far away from home, she put strong pressure on me to

---

[8] I am by no means implying that I would have been successful, even focusing on narrow, limited problems. I have colleagues who are as broadly educated as I, yet still find it possible to produce good mathematical research. On the rare occasions when my mathematical muse has inspired me, I have been just as narrow and lacking in perspective as anyone else.

stay with Northwestern, and that is how I came to go there. Despite the relative nearness of Northwestern, only a five-hour train ride from Alton, my mother was emotionally distraught when the day came for me to leave for college. The ten weeks from the start of classes to the Thanksgiving holiday were for her a nearly unbearable abyss of time. Three years later, when I informed her I would be going to graduate school at Princeton, her reaction was, "But if you go out there, you won't get home until Christmas!" I mention these things as an illustration of the different influences that family values have in different social classes. Families with a history of university education make the quality of the institution, not its geographical remoteness, the determining factor in choosing a university. (In saying that, I do not mean to imply that Penn is a better institution than Northwestern, only that the relative quality of the two places did not enter into the eventual decision.)

Undergraduate Life

Although social class was palpable and snobbery was rampant at Northwestern, where fraternities and sororities ruled the social life on campus, I remained blissfully unaware of any of it and could not have cared less about it. Although the family friend who had led me there was in a fraternity and urged me to go through Rush Week, I dropped out early on, out of sheer lack of interest, and spent the time instead hanging out with my new roommate. When I went back home, some of my high school friends asked me if I had joined a fraternity. When I said no, they asked me if I had received any bids. I cheerfully replied that I had not. Like my fiscally conservative father, I saw fraternity fees as a needless frivolity[9] and the social events the fraternities provided as boring distractions. I was by this time engrossed in intellectual activity and in my spare time preferred to eat pizza or go to see films with the other independents in my dormitory. I was, in short, a nerd.

From Physics to Mathematics

I took some physics courses at Northwestern, namely thermodynamics and modern physics. Neither of them meant much to me, since I did not have sufficient background in classical mechanics to appreciate the issues they dealt with. I had already learned, in high school, that I had no laboratory skills of any kind. My experiments, when they did not result in blasting a thistle tube into the ceiling, gave results that agreed very poorly with theory. They resembled, in fact, the shop projects I had completed in seventh grade, when the schools were still separating boys from girls and forcing the former to learn how to use chisels, drills, and hammers while the latter were supposed to master the functioning of ovens, electric ranges, and sewing machines.

After many decades of sporadic efforts to master physics, I finally came to the realization that it would never satisfy my obsessive pursuit of mathematical elegance and precision. Physicists, for example, are quite content to let a cubical box stand as the model when analyzing boundary-value problems for the three-dimensional wave equation. For me, that

---

[9] I was paying the full cost of college above the generous General Motors scholarship I received, working at the University Library during the semester and in a drug store during the summer. My parents made no financial contribution to my education at all.

approach would be an abdication of responsibility. I want to know what conditions lead to a unique solution, how general the boundary of the region can be, and how the solution can be expressed in terms of the initial boundary conditions. I almost never get through one chapter of a physics textbook without going off into numerous digressions involving pure mathematical problems, most of which I find myself unable to solve. Thus, I learn very little physics, and that not as physicists understand it.

In the end, I became a mathematician because the amount of mathematics I thought I needed to know in order to understand physics was more than I could learn as an undergraduate. In graduate school, I let go of physics and got started on the career that I eventually had. But graduate school also brought out the inherited proletarian and utilitarian values that I possessed. I shall begin my discussion of these issues with my experience at Princeton from 1963 to 1966.

Social Values in the American Mathematical Community

Even before I reached Princeton, serious questions had occurred to me about the value of the profession I was choosing. My father's jaundiced view of college educated engineers continued to echo in my head. While my fellow students at Northwestern, who were preparing for engineering careers, were doing co-op study, taking five years to graduate and interspersing terms in college with terms working in industry, my only employment had been in the library. I had sought more technical work during the summer, but discovered that I had no skills that were of any interest to the firms where I applied. I began to wonder whether mathematics was as useful as I had assumed it was in my uncritical acceptance of what my father had said about the importance of learning it.

During a conference at the office of Professor Art Simon, one of my classmates asked the question that was on both of our minds: Who pays a mathematician for doing this work? Simon pointed out that he himself had a career that paid modestly well. While that moved the issue back one notch, it immediately raised another issue in my mind: Why are people paid to teach something that has no intrinsic economic usefulness? This is not the kind of question that appeared to bother those of my classmates who came from more intellectual backgrounds than I did. For them, learning was a value in itself, and of course people should desire it and be willing to pay for it.

For me, however, the issue was more complicated than that. It would be one thing, I thought, to offer one's services as a teacher of mathematics to people willing to pay tuition to learn mathematics. In such a transaction, the question of the utility of what was being taught did not arise, unless that utility was misrepresented to the client. But I was being supported in graduate school by the National Science Foundation to undertake mathematical research; teaching was not even part of my duties as a graduate student. The government, through NSF, was purchasing my services as a future mathematician. I wondered why.

These were the days when the Soviet space achievements loomed large, and the United States was trying to "catch up" with their space program. That is why so much public money was being funneled through NSF and a comfortable portion was trickling down to me. The politicians who voted these expenditures must have believed (or would have believed if the issue had been presented to them) that my studying Banach algebras had some connection with the program to send people to the moon. I could see clearly that it didn't, and the discrepancy bothered me. But, of course, prudent economic self-interest dictated that I would stay in graduate school and continue to accept my fellowship. One of my classmates, after listening politely to my doubts, asked me why I remained in graduate school. Trying to rationalize, I told him that different government programs benefited different people and that I would, all my life, be paying taxes to support programs that benefited others. In this way, I got a sufficient sense of entitlement to soothe my conscience. My classmate, however, punctured that illusion, replying, "Well, tell me then, are you really interested in mathematics, or is this just your way of getting even with the government?" (Actually, it was a little of both, I now think.)

Throughout my career, one of my interests has been to look for genuine applications of complex mathematical ideas. Where I have found them, they have nearly always been models for thinking, useful as a concise summary of complex phenomena, more a guide to understanding and synthesis of what is known than a source of any specific knowledge that could not have been attained otherwise.[10]

Only recently has it become clear to me that theoretical understanding really is of great practical value. The instruments by which medical laboratories, for example, measure our cholesterol, blood sugar, and other vital signs, are not simple mechanical devices whose functioning is obvious to anyone who takes the trouble to disassemble them. The same is true of the components of the dashboard instruments in our automobiles. Their functioning, along with the numerical data they produce, is interpreted in accordance with abstract physical and chemical theories. If those theories are of no value, we ought not to believe our instruments; and GPS navigators are a waste of money.[11]

A Genuine Discovery Traceable to Pure Mathematics

In what I am about to relate, I rely on secondary sources. It may be subject to refutation by people who know the details of the actual history better than I. The discovery I have in mind is the invention of the radio (actually wireless telegraphy in the beginning). The realization by Maxwell in 1861 that light was an electromagnetic phenomenon and that its velocity was determined by the magnetic permeability and electric permittivity constants, led to experimental confirmation of electromagnetic radiation by Heinrich Hertz. Reading Hertz's obituary suggested the possibility of wireless telegraphy to Guglielmo Marconi. In this case, one can plausibly argue that pure mathematics in a physical theory led to a scientific discovery with practical applications, one that would not have been made, the

---

[10] Nowadays, I find myself in the same position in the public debate in the U.S.A. over evolution, a rear-guard battle still being fought by obscurantists using every rhetorical device they can foist on an ignorant public. It is clear to me that evolution is the only available scientific explanation that unifies the history of life on earth and that creationism is pure bunk. But when someone asks me why this difference is important and what practical contribution evolution makes to a student's understanding of biology or geology, I find it difficult to produce a reply that does not presume that a streamlined and efficient scientific theory has intrinsic value. That value is not recognized by the creationists, whose priorities tend to run in other directions.

[11] Obviously, they do function and give reliable results. But they had to be designed by people who understood the quantitative physical and chemical theories that are the basis for the numbers that appear in their output. It is inconceivable that they could have been cobbled together by practical-minded people doing Edison-style tinkering. Without theory, they would never have existed.

possibility of which would have been unrecognized—indeed it was unrecognized for more than 30 years—if the theoretical conclusion had not been known in advance.

Such applications are rare and represent only a minuscule fraction of the mathematical results that have flooded the journals for the last 200 years. I do not look for any comparable applications to arise out of point-set topology, to name just one example of the dozens that would fit. I therefore return to the theme of justifying the present scale of pure mathematical research.

The Educational Controversy over Mathematical Research. I was not the only one who noticed the impracticality of mathematics back in the mid-1960s. At the meeting of the American Association for the Advancement of Science in Montreal in December 1964, Richard Hamming (1915–1998), inventor of the famous Hamming codes, spoke out against the excessive abstraction of mathematics. When asked for an example of such, he cited the geometric line, visualized as having length but no thickness.[12] That was too much even for me, or perhaps I just felt a need to defend my chosen profession. I mentioned this controversy to Salomon Bochner, who later became my adviser. I observed (astutely, I thought) that surely Kepler's laws, a piece of applied mathematics that revolutionized astronomy, used precisely such abstractions, in the form of elliptical orbits. By Hamming's logic, it was Brahe's voluminous tables of planetary observations that represented "real" science, and Kepler was wasting his time. Bochner was more sanguine than I. He replied, "From time immemorial, there have been people trying to prescribe what mathematicians should do. But we are just like women in the world's oldest profession: they can't keep us from doing it."

This was a bon mot, to be sure. But while they can't keep us from doing it, they can refuse to pay us for it. The issue remained in my mind. Again, I was not the only one to notice and to wonder about the status and value of the mathematics profession. I had no worries about those mathematicians who worked in industry and commerce. They were surely providing a service of value to their employers. But academic mathematicians were another matter. Many were supported by various agencies of the government, including the military, until Senators Mansfield and Proxmire began raising serious doubts as to the wisdom of spending public money this way. Outside of support from government agencies and a small amount from private foundations, the majority of most research mathematicians' income is derived from teaching at universities. Here, I felt, a bait-and-switch was occurring. Students were charged tuition to cover their instruction, but a considerable portion of that revenue was being paid to professors for doing research that the students never even got to see, much less benefit from. Moreover, those professors were rewarded for success by further reductions in their teaching loads.

This aspect of mathematical research still troubles me today; but, as I have learned from many discussions, it does not trouble many of my colleagues. Even so, while working on this article, I found several bulletin boards on the web where non-mathematicians were discussing the usefulness or uselessness of abstract mathematics. Over the decades, others have addressed the issues I raise here. I mention in particular Morris Kline's *Why the Professor Can't Teach* (St. Martin's Press, 1977), and its generalization to other areas by Charles Sykes, *Profscam* (St. Martin's Press, 1989). I am aware that this issue—and its

---

[12] See the *New York Times*, December 30, 1964.

importance or lack of importance—is a matter of a person's values. It may well be more an indication of my own eccentricity than an item of concern to the community. I offer this concern in print so that readers can take it or leave it. When I have raised the issue in face-to-face discussions with other mathematicians, the response has often been, "Call me when you get over it." I admit that there is no universal moral imperative to share my concern. Still, I hope all mathematicians will remember that the economic well-being of our profession does depend on public support and that, to many in the public, university professors are the people who invented the fifty-minute hour, the three-day week, and the seven-month year.

At this point, I shall cease my *cri de coeur* and present the results of my informal investigation into the connection between social class and a mathematical career.

A Modest Statistical Experiment

In order to test my hypothesis that careers in mathematics are influenced by the socio-economic background of those who become mathematicians, I took a "random" sample of biographies of 85 mathematicians, which was unavoidably biased. The unavoidable bias came from the fact that these were all mathematicians of note.[13] The database from which I worked was the MacTutor website at St. Andrews University in Scotland, which has thousands of biographical sketches of mathematicians.[14] Although the sampling was random within the database, I was able to use only mathematicians whose biographies contained information on the family background of their subjects. This requirement added further bias to the sample, since the biographies of living mathematicians tended not to include such personal details. To add homogeneity to the sample and make some tentative conclusions possible, I restricted the sample to mathematicians born since 1800 in the United States, the British Commonwealth, or Western Europe.

For each of my 85 sample mathematicians, I recorded the date and place of birth and assigned a socio-economic class to the family based on the parents' (usually the father's) occupation. I had about a dozen such categories, including Self-Employed, Clergy, Military, Government, Professional, Teacher, Laborer, and Farmer. The category Self-Employed included all those who owned their own businesses, whether large or small: shopkeepers, merchants, wholesalers, and the like. The Professional category included university professors, lawyers, physicians, architects, and the like. The Teacher category included all teaching below the university level.

I put the data into a spreadsheet and looked for correlations. Of the 85 people in the sample, 62 were from just three countries: the USA (25), the UK (23), and Germany (14). By far the commonest family occupation was Self-Employed (24). It was followed by Professional (15) and Teacher (12), thereby accounting for 51 of the 85 people. The dates of birth were distributed rather evenly by quarter-centuries: 1800–1825 (5), 1826–1850 (13), 1851–1875 (17), 1876–1900 (16), 1901–1925 (22), 1926–1950 (12).

I had assumed that very few children of farmers would become mathematicians. Indeed, I had not expected any, since the only example that I knew of was Jakob Steiner. I thought

---

[13] Need I remark that my own name was not in the database from which the sample was taken?
[14] See www-groups.dcs.st-and.ac.uk/~history/.

there would be a slightly larger number of children of laborers who would select such a career. My data did not confirm these assumptions. I found five children of laborers among the 85 and six children of farmers. When I stratified the data, however, I noticed an interesting thing: All six of the farm children had been born in the nineteenth century; all five of the children of laborers had been born in the twentieth. I might explain these facts (if they are not a mere coincidence) by supposing that the two classes are probably hereditarily linked. The mass migration of people off the farm and into the city in the early twentieth century meant that the relatively small number of people from this socio-economic group who became mathematicians would be arising from the urban labor force rather than from the rural population.

The only other parental occupation that produced more than one or two mathematicians was that of the clergy. There were six children of clergymen among my sample, two each from the USA, the UK, and Germany. All but one of them were born in the nineteenth century. Since the clergy are particularly involved with abstractions that are not immediate objects of sense experience, I am not surprised to find their children inclined toward mathematics. Indeed, I have occasionally lectured on the analogy between mathematics and theology.[15]

A better-designed study would require surveying a random sample of all mathematicians, not merely those distinguished enough to merit a web page at a site devoted to the history of mathematics. It would also require a more formal definition of socio-economic class and demographic data on the proportion of the general population belonging to each class. Still, I think the data do show that the children of teachers, or people in the professions or in business for themselves, are more likely to become mathematicians than the children of laborers and farmers. To that extent, my study tends to confirm what I intuitively believed. As for the influence of socio-economic background on values, I doubt that anyone questions it in the abstract. Whether the specific examples I have given here can be explained by my own socio-economic background is unknown, even to me.

---

[15] What subject is queen of the sciences? Petrarch said it was theology. Gauss said it was mathematics.

14

The Badly Taught High School Calculus Lesson and the Mathematical Journey It Led Me To

Keith Devlin

Looking back, two things in particular led me to become a mathematician. The first was calculus, which I met when I entered the "Sixth Form" at age sixteen.[1]

The teacher introduced differential calculus to the class—there were about twenty of us, as I recall—in what, looking back, was probably a terrible approach for most of my fellow students. He began by pointing out that, given a real number x, we can raise it to a positive integer power n to give x^n. We can then do various things to the result: multiply by a constant a to give ax^n, raise it to another positive integer power m to give x^{nm}, extract its mth root to give $\sqrt[m]{x^n}$, etc. "Now here is something else we can do to x^n," he said. "You can multiply by the exponent and reduce the exponent by 1, to give nx^{n-1}."

The fact that I can remember this—almost verbatim to the best of my knowledge—almost fifty years later should indicate that this was a significant turning point in my life. What gave it that impact was what my teacher did next. After letting us do a few specific examples, $x^2 \to 2x$, $x^3 \to 3x^2$, etc., he told us that the resulting formula nx^{n-1} gives the slope of the curve $y = x^n$ at any point x.

I was stunned. Until that moment, I could not see any reason to consider this weird-looking operation. Now, suddenly, he had dropped a bombshell. As far as I remember (and parts of this are less clear in my mind), he went on to explain to the class that curves have a continuously changing slope,[2] and that in order to compute the slope at a given point you have to find the formula that describes the slope and then plug in the x-coordinate of the given point. In the case of $y = x^n$, that magic formula is nx^{n-1}. I got at once the point about the slope itself being a function. But what flabbergasted me was the formula itself. It looked

---

[1] I was born and grew up in Hull, England. In the British system at the time, mandatory high school finished with the "Fifth Form" at age sixteen. Pupils who intended to go on to higher education—about 3% of the school population—remained on for two further years in what was called the Sixth Form, where they took just three subjects (of the student's choosing).

[2] All curves were continuous in the Sixth Form, indeed given by an algebraic or analytic function.

so bizarre. Simple, to be sure, but seemingly arbitrary. Why that sequence of operations? Who had come up with that formula and how?

As I worked my way through a number of simple examples the teacher gave us to compute slopes of curve at given points, my mind was racing to make sense of it all. But I could not. It made *no* sense. It was a magical, rabbit-out-of-a-hat move. Yet it gave the right answer.

The teacher told us that, as the term progressed, we would learn more formulas for slopes of curves. The process of going from x^n to nx^{n-1} was just one of many cases of a process called "differentiation."

Until that moment, I had intended to go on to university to study physics. That goal had been fixed in my mind since 1957, the year before I went to high school, when the Russians launched Sputnik, the first human-made object to orbit the earth. Inspired by the dawn of the space age, I'd learned a lot of physics, certainly enough to appreciate the importance of computing slopes of continuous curves.[3] But I was so perplexed by that strange formula—and who knows what other surprising formulas for derivatives we would learn in the coming months—and so stunned by the the fact that this strange-looking process called differentiation gave something so incredibly useful, I knew I could not rest until I understood what was going on.

A few weeks later, the teacher revealed the secret to the x^n magic, essentially showing us the standard derivation of the derivative of that function (absent ϵ, δ rigor), in particular, how that term nx^{n-1} comes from an application of the binomial theorem. Now it was no longer a mystery. But, in place of that strange symbolic magic, was something that I found even more impressive: the fact that someone had been able to figure this out in the first place.

Until that moment, mathematics had been for me a sequence of fairly predictable, easily understood techniques for solving certain kinds of problem. Knowing that I needed to master mathematics in order to become a physicist, I had put effort into the subject (though not to the extent of doing a whole book-load of problems over the summer) and become quite good at it. But it had never grabbed me. There was no mystery, no magic. Nothing that suggested human creativity.[4] Now everything had changed. I was intrigued, and I wanted in on the secret—on all the secrets, for we knew there were more to come.

As I hinted at earlier, I suspect my reaction to the teacher's introduction of the derivative was somewhat different from that of my classmates. My recollection is that they were pretty baffled by this strange new operation, and certainly had little curiosity as to where it had come from. "From Chapter 11," I can imagine being their answer to that question, though I have no recollection of anyone actually asking it, let alone giving an answer.

---

[3] At the end of my fifth year at high school, the physics teacher gave the future Sixth Form physics students the textbook we would be using the following year, suggesting we might glance through it over the six-week summer break so that we had some idea what was in store for us. I got hooked immediately. When the school reconvened in September, I handed the teacher a fairly thick pile of neatly handwritten work. I had gone through the entire book and solved every single problem. I'm told my teacher recounted that story to every physics class he had in the years that followed, long after I had left the school. I'm not sure whether that had a positive or negative impact on the students he told the story to.

[4] Only many years later did I come to appreciate that numbers and elementary arithmetic were equally impressive feats of creativity, but since the creative work was completed many centuries ago and we are taught about them from a very early age, we accept them as a mundane feature of life.

On the whole, I do not recommend introducing differential calculus the way my teacher did. It worked for me because (1) as an intending physicist, I could see at once the importance of a mathematical formula that gives you the slope function of a given function, and (2) since as early as I can remember, I have always wanted to know how and why things worked.

Three Little Dots

The derivation of the derivative intrigued me in part because it involved a thought process that went beyond the finite limitations of the human mind. It was possible to determine the limiting value of an expression like $[f(x + h) - f(x)]/h$ as h approaches 0, without actually setting $h = 0$, a value that renders the expression meaningless.

Going beyond human limitations, this time the limitations of language, were at the heart of the other event that made me decide to pursue mathematics as a career.

After my calculus epiphany, I started to devour every "popular account" of mathematics I could lay my hands on. There were not many back then. I began with Lancelot Hogben's classic *Mathematics for the Millions*, lent to me by one of my two math teachers (one for pure mathematics, the other for applied, a common arrangement back then), and then acquired—by asking for them as birthday or Christmas presents—paperback editions of the books *Mathematician's Delight* and *Prelude to Mathematics*, written by W.W. Sawyer.[5]

At that point, my two math teachers decided that I was about to become a disruptive, and likely discouraging, influence on the rest of the class. Accordingly, they arranged that I would henceforth be excused many of the regular math classes, and instead sit in the corner of the school dining hall teaching myself from their college textbooks, which they brought in for me. There, I would study alone, uninterrupted apart from an occasional visit from the teacher to see if I needed any help, and the dining hall staff who wanted to put out the salt and pepper.

As I worked my way though the various books, I kept coming across expressions such as

$$a_1, a_2, \ldots, a_n$$

I remember being puzzled at first. What kind of a mathematical operation did these three dots denote? I had already learned that "real mathematicians" use a dot to denote multiplication (the cross was for the kindergarten), and I had encountered Newton's dot notation $\dot{r}$ for the derivative of r with respect to t. But three dots in a row? I looked, but nowhere in any of the books I had at my disposal could I find an explanation of what algebraic operation those three dots denoted. I could, of course, have asked one of my teachers, but since the notation was ubiquitous, yet never explained, it was clear to me that it must be something so common and so obvious that I would show myself as a complete imbecile if I were to admit I had somehow never learned what this mysterious mathematical operation was. Perhaps my mind had been wandering when the teacher told the entire class what it meant.

Eventually, by trying to make sense of what was being said whenever these three little dots appeared, it dawned on me. It wasn't some mysterious algebraic operation. It was just

---

[5] My highly positive, transformative experiences with popular accounts of mathematics are what led me to devote some of my later career to writing similar books.

a special use of the familiar ellipsis of ordinary language. It meant "continue the sequence in exactly the same way until you get to the final member." The reader was being instructed to recognize the pattern established by the first few terms given, and imagine it continuing for possibly millions of steps. Or, and now I was really blown away, I would encounter an expression such as

$$a_1, a_2, a_3, \ldots$$

where the writer was telling me to imagine the pattern to continue *forever*.[6]

Looking back, my early experiences in (advanced) mathematics in the high school prepared me for the moment many years later, in 1988, when I read Lynn Arthur Steen's article in *Science*, titled "The Science of Patterns." Steen's thesis really struck a chord with me, and a few years later, in 1994, I wrote my own essay on the same theme, published as the book *Mathematics: The Science of Patterns*.

What Is Mathematics?

For me, mathematics—perhaps I should say the mathematics I love—has always been the science of patterns. But that's just me. I have learned over the years that we mathematicians see and approach our subject in different ways. For some, the thrill is in the chase. They see math primarily as an endless series of challenges, as problems to be solved. Paul Erdős was a well-known example of this ilk. Others, and I include myself in this category, are attracted more by the big picture. The overall landscape is more important than scaling any one peak. We mathematicians of this second stripe are, to use common descriptions, theory-builders rather than problem-solvers. To be sure, this is a simplistic division, and I doubt any mathematician lies totally in one of those two camps. Most, if not all, of us get pleasure from both solving a problem and from the elegance of the big picture.

The fact that I have written twenty-eight books (at the time of writing this essay) reflects my strong preference for big picture synthesis. Roughly a third of those books are research monographs, so even when engaged in original research I am attracted to the development of a comprehensive study or an overall theory. Many of my more expository books, where I describe the work of others, are an attempt to bring out the unity, the hidden connections, and the overall beauty of mathematics. In recent years, when my research went off into decidedly applied, new areas of mathematics—in some cases so applied that the use of the term "mathematics" becomes questionable (linguistics, communication, human action, and intelligence analysis)—I have devoted much more effort to expository writing about mainstream mathematics than earlier in my career. I suspect this is to fulfil in me a need to continually experience the beauty and structure of mathematics, and to share it with others. For my new areas of research are for the most part messy, multidisciplinary, and open-ended, rarely showing any elegance, and clearly reflect another part of my brain than the one that initially fell in love with mathematics.

---

[6] Later on in my mathematical career, as a graduate student in set theory and infinitary combinatorics, I learned to extend the idea into the transfinite, with über-sequences such as $a_1, a_2, a_3, \ldots, a_\omega, a_{\omega+1}, \ldots, a_{\omega+\omega}, \ldots, a_{\omega \times \omega}, \ldots, a_{\omega^\omega}$.

Along with this shift in my research, including a growing interest in the equally messy and open-ended area of K-12 mathematics education, my overall philosophical view of what mathematics is has changed. (I am not sure if there is any causality here, and if so in what direction.)

No Longer a Platonist

I began my professional mathematical life as an out-and-out Platonist (i.e., one who believes that ideas have a separate existence outside of our heads), believing that the objects of mathematics exist in some abstract realm that is independent of humankind. What mathematicians do, I believed, is discover eternal truths about those objects. I now see my subject differently, as something we create, a product of the human mind.

Like any form of creativity, the mathematical variety is subject to certain restrictions. For example, architects have a lot of freedom in designing a building, but it has to have structural integrity, it must stand up, last for many years, resist the ravages of climate and use, and it must fulfil the purposes for which it is built. This means that, while two buildings may appear quite different, they will have to share a number of common features. So too with mathematics, only the restrictions that govern mathematics seem far more strict, and restrictive, than any other discipline I know of.

The nature of mathematics—of what constitutes mathematics—means that, once a mathematical object or structure has been defined, it is (mathematically) the same for everyone, and further investigation of that object or structure will, to all intents and purposes, be a process of discovery.

The only place where there is (at least in principle) unfettered creative freedom is in the initial choice of the postulates or axioms that start a particular mathematical ball rolling. Here, however, logical coherence and utility (in the real world or elsewhere in mathematics) act as demanding and unforgiving filters. Since so few of these creative acts can pass through this filter—indeed, most mathematicians have good instincts that prevent them heading down an unprofitable path in the first place—this adds to the illusion that the few notions that survive somehow existed already. Hence the Platonistic philosophy.

Though I myself am no longer a Platonist, many mathematicians are. Both Platonism and its negation seem totally harmless beliefs that do not affect mathematics itself.[7]

Doing Mathematics

How does a mathematician—by which I mean a person who (a) calls herself or himself such and (b) whom other mathematicians agree is a mathematician—do mathematics?

Well, I can point to various activities that are part of the process, such as searching for an underlying, abstract, structural/logical pattern, making guesses, looking for logical consequences, looking for logical causes, asking questions about the nature of or connections between mathematical objects or structures, deriving arguments to establish the truth of various statements about mathematical objects or structures, and more.

---

[7] I think there may be evolutionary-neurophysiological reasons why *doing* mathematics tends to entail a Platonistic belief, an idea I investigated in my paper [1], but will not pursue here.

If the focus of these activities comprises mathematical abstractions, then what is being done is often referred to as "pure mathematics;" if those same processes are carried out for various features of the real world, "applied mathematics" is the common description. Personally I find this a highly artificial distinction that can be defended only in terms of making political and social decisions, including the allocation of space and funding.

But can I say more about what it means to do mathematics? Not much. Not even for myself, let alone anyone else. Occasionally, when I'm trying to understand some piece of mathematics that is new to me, or am struggling to synthesize a new theory, or am wrestling with a mathematical problem, I become aware that I am "doing mathematics," but then that very act of self-awareness interrupts the process and I find I am no longer actually doing math. And as for those genuinely creative moments when everything falls into place or a problem is solved, like every other mathematician who has been motivated to try to articulate the process, I have to admit total defeat. Those breakthroughs just happen, often at an unexpected moment when my mind is focused on something else.

When I do mathematics, do I think in words, in symbols, in images? I often get asked this question, usually by non-mathematicians. For sure, if they were to watch me at work, either alone or collaborating with one or more colleagues, they would see me write (and maybe utter) words (rarely complete sentences), scribble down symbolic expressions, or sketch simple diagrams. But none of those are at the heart of the process.

The best way I can describe the sensation of doing mathematics is that it is living in a virtual world. These days, that term conjures up digital environments such as *Second Life* or the popular videogame *World of Warcraft*. But before we had those technologies, novelists regularly created virtual worlds for us to experience, and that is closer to the virtual worlds that constitute the different domains of mathematics. Mathematical domains are determined by linguistic descriptions, descriptions written in words, symbols, and diagrams.

The person reading a novel rapidly ceases to be aware of the words, and becomes a constituent (generally a passive, observing constituent) of the world the novelist creates, thereafter *experiencing* that world and *discovering* what its properties are and what actions unfold therein. Likewise, the mathematician works by entering the relevant virtual world, and from then on the process is one of *self-directed, active, experiential discovery*. Words, symbols, and diagrams no longer have much significance. They were there to help establish that world. The mathematician may resort to those linguistic tools to record things suspected or discovered, or to help plan out future actions in the world, but that is really no more mathematical discovery than when a geographic explorer makes notations on a map.

All of us become familiar with the house in which we live and the neighborhood in which we go about our daily lives. That is the relationship I had with set theory, the domain of mathematics I worked in for the first part of my career. The world of set theory felt real to me. I knew how to find my way around. It had existence. I experienced it. My research work consisted of exploring the parts of the world I did not yet know about. Over the years, I had built up a body of intuitions that could guide me when I wandered into new parts of the world.

Trying to describe that process in terms of my using words, symbols, or pictures would be totally misrepresentative. Of course, I had to use all three to present my work to others. But, for that communication to succeed, those others would have to use my words, my symbols, and my diagrams, as guides to their own journey through the same world. What I did when I was *doing the math* was live in and explore that world. Again, I think the analogy with writing and reading a novel is pretty good here—particularly a detective novel.

Incidentally, I am pretty sure that all other mathematicians do mathematics in the same way (though some may choose other ways to describe it), and I am equally sure that this is why we mathematicians are so often described as being "unworldly." When we are deeply engaged in mathematical thought, we literally are cognitively living in another world.

The popular image of mathematics as being rule-based, rigid, and uncreative is about as far from the reality as can be imagined. I think this perception is so common because very few people get beyond the words and the symbols and the rules. If you have not mastered the words and the grammar of English language, you are not going to be able to enjoy a novel. Your experience of reading a book will be of trying to *read*. You will never enter the world the novelist is trying to create for you, and never experience that world and what takes place in it. The same is true of mathematics. For many people, they never got beyond learning the words and the grammar—the symbols and the rules. The world those symbols and rules create remains forever inaccessible to them. Sadly, many people are not even aware of what they are missing; they think that what they experienced *is* mathematics. They may say they could not do math or that they did not like it, but in reality they never got there!

Let me say a bit more along this line. My focus in this essay is advanced mathematics. That, after all, is what (professional) mathematicians do—and the psychology of the professional mathematician is what this volume is about. The farthest most people get into mathematics is a bit of elementary arithmetic and geometry.[8] Yet some of us do manage to get further. How is that possible? Well, in this case, the real world we physically live in provides the relevant experiential domain in which those more elementary parts of mathematics can be done. The more a young child can take the arithmetic (and geometric) parts of the physical world and convert them into a mathematically-efficient virtual counterpart-world, the better she or he will become at solving math problems. This is why some children seem to make math look "easy." Once they have that virtual world available to them, it *is* much easier. The world does much of the work for them; they just have to negotiate and explore it.

The Secrets of Success

Doing mathematics is not a process of staggering about mindlessly and blindly in this virtual world I have talked about. When I say that doing mathematics is a process of exploration, then like all explorers—at least the successful ones—we adopt strategies to improve the likelihood of solving the problem we set out to do or discovering something of value.

In other words, mathematicians learn *how to set about solving a problem*. The techniques we use have been written about extensively: draw a diagram (if applicable), look at some specific examples, try simple cases, see if there is a meaningful generalization, look at solutions to similar problems, try changing the question, try to prove the negation, look

---

[8] From the perspective of what mathematicians mean by algebra, "school algebra" is essentially arithmetic looked at from a different perspective, though that shift in perspective is a significant one that many students fail to make.

for analogies with other parts of mathematics, talk about the problem with as many other mathematicians as you can, etc. But it is rare for any of these activities to culminate directly in a solution. Rather, they *lay the groundwork*. They *put in place conditions that make a breakthrough possible*.

From what I have said, it should be clear that, when a mathematician starts to work in a new domain, or sometimes even on a new problem in a familiar domain, she or he has to put in considerable effort to create that deep understanding and familiarity I am calling a virtual world. That takes time and effort. The focus is on the domain or the problem, but for a long time (long could be a few minutes, a few hours, a few days, or even months or years, depending on the circumstances) the only real payoff is an increased understanding—establishing a richer and better understood virtual world, to continue with my analogy. Even when that stage could be said to be complete, however, success may be a long time coming, if at all. Which leads me to another important point.

What does it take to be successful in this enterprise? (Success is a relative term, so you have to be careful in interpreting my remarks. If the metric were proving Fermat's last theorem or solving one of the Millennium Problems, hardly any of us would be successful. Fortunately, there are less stratospheric levels of success that are still perfectly meaningful to the individual concerned.)

What it takes is perseverance. Sure, talent plays a role, knowledge plays a role, brilliance plays a role; so too can luck. But to my mind, the most significant factor is not wanting to give up. It won't work every time. But if you give up, then for sure you won't solve the problem.

I have found that if I keep plugging away at a particular problem then, on occasion, I will eventually find the solution. Not always, by any means. In fact, truth be told, more often not, though on a couple of occasions I have given up after a long struggle only for someone else to succeed by proceeding exactly the same way and making one ever-so-tiny additional step that I missed. Those moments are extremely frustrating, but fortunately are rare. Much more common is when someone else solves the problem and I think, "I would never have got that."

To get back to my main point, however, there is one particular factor that results in success in mathematics, and that factor is persistence.

Needless to say, prerequisites for persistence are a deep interest in the domain and actually enjoying the process. In other words, to be a mathematician you need to enjoy the landscape in the mathematical universe and you have to have a penchant for exploration, sometimes requiring considerable ingenuity and creativity.

You also need a tolerance of repeated frustration and failure. For most of your explorative episodes will turn out not to lead you to your destination. It helps if you nevertheless enjoy those excursions.

The Mathematical World I Entered

As I mentioned earlier, for most people, their only acquaintance with mathematics was during their school years. That means they have seen little or no mathematics that is not at least several hundred years old. That is particularly unfortunate, since the very nature of the subject changed during the nineteenth century. One change was in geometry, with the discovery of non-Euclidean geometries. This development is fairly well known outside the field, though what is less known is that those discoveries led eventually to a different conception of what geometry actually is.

Putting geometry to one side, however, there was another major shift in the way mathematicians saw their subject, that is hardly known at all to nonmathematicians.

Up to about 150 years ago, although mathematicians had expanded the realm of objects they studied beyond numbers and algebraic symbols for numbers, they still regarded mathematics as primarily about *calculation*. Proficiency at mathematics depended above all at being able to carry out calculations or manipulate symbolic expressions to solve problems. (Remember, I am putting geometry to one side here.)

In the middle of the nineteenth century, however, a revolution took place. In the small university town of Göttingen in Germany, the mathematicians Lejeune Dirichlet, Richard Dedekind, and Bernhard Riemann led the development of a new conception of the subject. For these mathematicians, and others, the primary focus was not performing a calculation or computing an answer, but formulating and understanding abstract concepts and relationships. This was a shift in emphasis from *calculating* to *understanding*. Within a generation, this revolution completely changed the way pure mathematicians thought of their subject.

For the Göttingen revolutionaries, mathematics was about "Thinking in concepts" (*Denken in Begriffen*). Mathematical objects were no longer thought of as given primarily by formulas, but rather as carriers of conceptual properties. Proving was no longer a matter of transforming terms in accordance with rules, but a process of logical deduction from concepts.[9]

Like most revolutions, the Göttingen one had its origins long before the main protagonists came on the scene. The ancient Greeks had certainly shown an interest in mathematics as a conceptual endeavor, not just calculation, and in the seventeenth century, Gottfried Leibniz thought deeply about both approaches. But for the most part, until the Göttingen revolution, mathematics was viewed primarily as a collection of procedures for solving problems.[10] To today's mathematicians, however, brought up entirely with the post-Göttingen conception of mathematics, what in the nineteenth century was a revolution is simply taken to be what mathematics is. The revolution may have been quiet, and to a large extent forgotten, but it was complete and far reaching.

The change in the overall conception of mathematics was accompanied by a rapid increase in abstraction. Finding themselves dealing with abstract entities that were not firmly rooted in everyday experience, mathematicians had to rely on formal definitions of concepts and rigorous proofs of results. It was not long before they began to encounter results that were counter to their intuitions. Their formal definitions produced strange objects, such as continuous functions (from real numbers to real numbers) that are nowhere differentiable, or continuous curves that completely fill the two-dimensional plane.

With clear evidence that their intuitions were no longer reliable—evidence from the developments in geometry, the discovery of the "monster" functions, and other surprising

---

[9] This new approach to mathematics now permeates all university mathematics instruction beyond calculus.
[10] That was true to some extent even for geometry.

consequences of their definitions—mathematicians began to analyze their concepts, and placed greater emphasis on clear specification of the assumptions (the axioms and rules of deduction) that underlay mathematics. In particular, what was essentially a new area of mathematics came into being, known as "Foundations of Mathematics."

By the time I was awarded my bachelors degree in mathematics from Kings College London in 1968, "Foundations" (as it was generally referred to) had become a collection of four, well-established, related branches of mathematics—logic (or proof theory), set theory, model theory, and recursion theory—and that was what I pursued in graduate school.

The Foundations of Mathematics

The name "foundations" was potentially misleading from the start. The word most typically conjures up the idea of a solid base on which something is built. In this case, however, the entire edifice of mathematics had been developed over several thousand years, had been successful, and was for the most part well understood. The detailed study of formal logic and the analysis of mathematical concepts such as number, function, continuity, etc. were foundational in an abstract sense of logical dependency, but not in terms of mathematical practice, which continued virtually unperturbed, while a relatively small number of mathematicians examined the subject's "foundations."

By the middle of the twentieth century, the foundational enterprise was more or less completed—successfully. Mathematicians had formulated axioms for the different number systems, had provided rigorous definitions of subtle notions such as continuity, differentiability, and integrability, and had shown how to construct all of the objects of mathematics out of pure sets.

In a strict logical sense, all of mathematics had been reduced to set theory, the branch of foundational studies that I focused on in my doctoral research at the University of Bristol. The reduction—it would be more accurate to describe it as demonstrated *reducibility*—showed how, for instance, the natural numbers could be constructed from "pure sets" (i.e., sets whose members are themselves sets, likewise for those member-sets, and so on, so that there was nothing that was not a set), and then how the integers, the rationals, the reals, and the complex numbers could each be constructed (set-theoretically) from the previous number-system in the sequence. Analogous reductions to set theory were provided for all the other entities and systems of mathematics.

For the most part, everyone involved in this process knew full well that natural numbers are not *really* equivalence classes of finite sets under equipollence, that rational numbers are not *really* ordered pairs of integers, and real numbers are not *really* ordered pairs of sets of rationals (Dedekind cuts), etc. Rather the process was one of providing set-theoretic *models* of the various systems in order to establish consistency. But on occasion this perspective was overlooked, and continues to be to this day. Presented with a well-constructed model, it is easy to confuse the model with what is being modeled. For instance, I have heard people make claims such as "The real numbers are not a subset of the complex numbers." This is precisely to confuse complex numbers with a model of the complex numbers in terms of the reals.

I was certainly not immune to making such an error, and likely am still susceptible. Not for number systems, but early in my professorial career, I would present my students with a *definition* of a "function" as a set of ordered pairs satisfying the property that there cannot be two pairs in the set that have the same first member but different second members. Looking back, I think that was an absurd thing to say. (Except when I was lecturing on set theory, where that definition was appropriate.) Mathematics is a way of thinking about that world that humans have developed over many centuries, and which has proved to be incredibly productive. That way of thinking has led to the development of many abstract notions that correspond to things in our world, to ways we think, and to things we do in the world. Functions are one of those notions. As such, functions are not sets, they are rules that associate. Sure, we can be very generous as to what constitutes a rule; it certainly does not have to be something written down in a language, formal or otherwise. But to view it as a set is a category mistake, confusing a model with that which is being modeled.

In the case of functions, I think this is a relatively minor issue. The very property that makes a function a function lends it to be modeled as a set of ordered pairs, and that model reflects its nature as a function. Not so in the case of number systems, where an important property is that $\mathbb{N} \subset \mathbb{Z} \subset \mathbb{Q} \subset \mathbb{R}$. Still, if you take the abstractions of mathematics to be formalized reifications of our intuitions, then a function ought to be a rule not a set. (Unless you are actually doing set theory!)

It is possible to take the reductionist program of constructing models as an *inspiration* for a synthetic approach to mathematics, without falling into the above traps. For a period, a group of French mathematicians headed up an international movement called the Bourbaki School that took just such a sophisticated approach. The Bourbakians sought to *develop* all of mathematics in a strict axiomatic fashion, beginning with "pure sets" or some equivalent starting point. Early in my career, I was very attracted by that approach, but later my enthusiasm waned. Today, the Bourbaki initiative has pretty well fizzled out, although it definitely had an influence on the way mathematics is now presented in textbooks and lectures.

Mathematics: Something We Know or Something We Do?

My reason for the above foray into the nature of "foundational studies" was to set the stage to describe my present conception of what mathematics is and what it means to do mathematics. As I have already indicated, or at least hinted at (more than once), my position has shifted over the forty-two years of my career since I received my Bachelors Degree in the subject.

I've already mentioned my shift from Platonist to "social constructivist." This shift is not at all uncommon among mathematicians as they progress through their careers. (Wittgenstein is a famous example that comes to mind.) But I have experienced another shift, and that is what I want to concentrate on now as I bring this essay to a close.

When I was a teenager reading those popular expositions of mathematics, the definition of the subject as "the science of patterns" was not yet in vogue. (Though it had been proposed.) One definition that I did come across—I no longer recall exactly where—is that "Mathematics is what mathematicians do for a living." I thought at the time it was a joke, and in part it surely was. But now I read much more into that assertion. (Whether or not the writer intended such, I don't know.)

Today, I think of mathematics as a body of knowledge that results from *mathematical thinking*. Mathematical thinking is a particular form of thinking, either about the world or about abstractions humans have recursively generated in the course of, or as a result of, thinking about the world. Like pornography, mathematical thinking is easily identified[11] but defies a simple definition. It includes strictly logical thinking, but is not restricted to that, and in fact goes well beyond it, though the conclusions reached by the process should be capable of logical verification. Not all mathematical thinking generates mathematics (i.e., mathematical knowledge); in fact most does not, being directed at practical problems that arise in industry, commerce, finance, and many other areas of human activity.

Though elements of mathematical thinking appear to arise naturally, and may be innate, for the most part it is a special kind of thinking that requires training and practice. Many people find some aspects of mathematical thinking counterintuitive and hard to comprehend; for example, proof by contradiction or the standard definition of continuity.

Over many centuries, mathematical thinking has given rise to a rich ontology of concepts and a substantial and impressive body of knowledge about those concepts, and that is what we call "mathematics."

Notice that this viewpoint is not Platonistic. It views mathematics as an abstract ontology and a body of knowledge about that ontology, that is derivative of mathematical thinking—a human, cognitive activity.

For me, this shift in viewpoint was likely in part occasioned, or at least encouraged, by my collaborating on research projects outside mathematics (often by invitation, as my writings and growing professional network made my work known to others outside the field). In working with experts from disciplines other than mathematics, however, I found that my contribution rarely consisted of my taking some known mathematical technique off the shelf and applying it. In most cases, they could do that themselves, and generally did. Rather it required my *thinking about the target domain or problem from a mathematical perspective*. What I could contribute, in other words, turned out far more often to be mathematical thinking rather than an application of some (known) mathematics.

Another stimulus for my shift, I am sure, came from my learning more of the history of the subject, which is something else that I have done later in my career. I used to think that mathematical knowledge was like piles of sand, that we add to as time progresses. Not so. Old sand is continually thrown away from each pile, to be replaced by entirely new sand, as the piles are re-interpreted. To give just one example, today we look at Newton's definition of the derivative and we see it as applying to functions from real numbers to real numbers. But that cannot possibly be what Newton was thinking about, since the real number system was not developed adequately until two centuries later. Rather Newton was focusing on physical entities he called "quantities." The identification of quantities with real numbers that occurred in the nineteenth century literally changed the very conception of what calculus is about, even though the mechanics of the method remained largely unchanged. If you want to see a steady *growth* in the development of mathematics, you need to look at mathematical thinking, not mathematical knowledge.

---

[11] At least by mathematicians, in the case of mathematical thinking.

Misleading Descriptions of Mathematics

The late Paul Erdős used to say "A mathematician is a machine for turning coffee into theorems."[12]

Another common description says that "The job of the mathematician is to establish mathematical truths (i.e., to prove theorems)." When challenged, the person making this statement can usually be persuaded to amend the definition to allow for the (occasional) crafting of a new definition or the formulation of axioms. But that is about as far as it goes.

I was comfortable with both statements for the first twenty years of my mathematical career, when both described pretty accurately what I was doing. (Well, accurately if you ignore that fact that I spent a great deal of my time teaching mathematics to university students, and since my contract said I was supposed to spend equal amounts of time teaching and pursuing research, it seems clear that my employers had a clear idea of what a mathematics professor is supposed to be doing.)

But those descriptions apply almost exclusively to individuals who are employed by research universities.

Moroever, they apply only to university mathematicians who are active in research. Now, I have no hard data, but based on my experience, only a relatively small percentage of university professors of mathematics are active in research, and the batting average for successfully proving theorems in mathematics is notoriously low. So, when it comes down to it, the vast majority of people who could justifiably describe themselves as university mathematicians are not at all engaged in the activity described by those crisp statements.

And then there are the many thousands of people around the world employed to do mathematics in industry, commerce, governments, and so on. They rarely if at all prove theorems, rather they use mathematics to solve problems, most of which would not qualify as "mathematical results."

Now, you are perfectly entitled to take a narrow definition of "mathematician," so that it applies exclusively to the Erdős machines. For many years, the International Mathematical Union published a *Directory of Mathematicians* that adopted just such a definition. Perhaps the editors of the volume you have before you implicitly had just such a definition in mind, though if they did they did not convey it to contributors (at least in my case). And perhaps it was that definition, with its hint of an intellectual "To boldly go where none have gone before" philosophy, that prompted you to pick the book up and start reading.

Having clearly nailed my colors to the post declaring that I (now) believe mathematical thinking is primary, with mathematical knowledge derivative on it, it will, I think, be obvious that I am going to argue for an inclusive conception of what a "mathematician" is and does.

When you do that, the "psychology of the mathematician" does not seem particularly different from the psychology of any other thinking, reflective person who seeks justifiable,

---

[12] I actually experienced that first-hand. Having solved an Erdős problem in my doctoral dissertation, I spent a brief period with him and his colleagues in Budapest, drinking strong black espresso and trying to solve problems. If I had known that Erdős numbers were going to become a mathematical "status symbol," I would have made sure we actually completed and published the brief joint paper we started. Since I did not, my Erdős number remains a respectable but fairly common 2, not the much-coveted 1.

rational answers to life's problems. To be sure, the modes of thinking in mathematics are, when aggregated, uniquely characteristic of the discipline. But that is just implementation.

I suspect the reason why many people view mathematicians as somehow cut from different cognitive or psychological cloth is purely an interface issue. You don't have to be a physicist or biologist or geologist or sociologist or psychologist or electrical engineer, etc. in order to have a sense that you know what they do, and to appreciate why they do it. That sense may be highly inaccurate, and in many cases probably is, but that does not prevent people from thinking they know what is must be like to be one of those professionals.

For mathematics, however, this is not the case. That impenetrable, symbolic/methodological mathematics–human interface means that, in order to get any sense of what mathematicians do, and how they think, you have to go at least part way to becoming one.

If you are an outsider reading this essay, then I suspect I have not come close to giving you a sense of what makes these "psychologically unusual" beings called mathematicians tick. I never tried. Instead, I set out to convince you that we are in fact no different from any other reflective person, yourself included. We just like math. And over the years we have learned that, while it is never easy (a mathematician who finds math easy is not working on the right problems), it is just about doable.

Reference

[1] Devlin, K, A mathematician reflects on the useful and reliable illusion of reality in mathematics, *Erkenntnis*, vol. 68, no. 3, May 2008, pp. 359–379.

15

The Psychology of Being a Mathematician

Sol Garfunkel

I would like to refine the topic slightly. For the past thirty years I honestly don't think that I could in fairness call myself a mathematician. I am and have been a mathematics educator, certainly from the early 1980s on. The psychology is subtly different and I'd like to take the time here to talk about those differences. I will attempt to be as honest and frank as possible and to distinguish value judgments from fact or theory.

First, my father, Jack Garfunkel, was a mathematician. Some of you may know of his work in geometric inequalities. He even has one named for him. My dad was typical of a certain generation. He was the top mathematics student at City College in 1930 and won the Belden Prize there. He longed to go on for his Ph.D., but it was 1930 after all and he wound up working in his father's grocery store on the lower east side of Manhattan. The grocery business led to the candy business in which he had a rags to riches (in 1947 he had the exclusive rights to Walt Disney chocolate molds) to rags ride up until 1960, when he took an emergency exam to become a high school mathematics teacher.

I have a number of his candy cookbooks, ones in which there are recipes for marshmallow and jellies where the ingredients are measured in bushels and barrels. And, in the margin of those books, are twenty-page proofs of Morley's theorem using trigonometric identities and inequalities. The point is he did mathematics every day of his life, despite the fact that his formal learning of the subject ended in his senior year in college. His Erdős number is one lower than mine and he was a regular and prolific contributor to *Crux Mathematicorum* and to the problem sections of the *American Mathematical Monthly*.

My father wanted me to become a mathematician and college professor. The words themselves cannot begin to express how strongly he wanted these things. He loved mathematics and he wanted me to love it as well. He had lived through the ups and downs of the business cycle and he wanted me to be spared its vicissitudes and uncertainties. He saw the life of a research mathematician as an ideal—his ideal—and he wanted that life for me.

And so he taught me mathematics. We did math every day that I can remember. As I got older we worked on the sorts of problems that interested him, mostly in geometry. He was in effect my lifelong major professor. Solving a problem, especially a particularly difficult one, was a way of gaining parental acceptance and praise. There was no better motivation. Speaking of geometry, my father loved to make sketches. He would work tirelessly on conjectures—drawing and redrawing configurations to test a potential theorem or inequality, in the days before geometric utility software. The most appreciated gift that I ever gave him was a highly precisioned draftsman's compass. There were many many times when he would use it to convince himself of a fact, but not have the ability to prove or disprove it.

This willingness to test out a hypothesis with lots of special cases has always stuck with me, both in my mathematical and personal life. My father loved a good problem or conjecture, whether or not he could come up with a solution. And in many ways that became my own strength, posing interesting problems and identifying problems that others would find interesting. I have good taste. Like my father, my technical skills were never my real strength; but like him as well I was not daunted by this fact.

My father's younger brother had a history of nervous breakdowns. In each case they were brought on by his inability to cope with demands placed upon him, beginning in college with all of the homework assigned. My father was tremendously affected by my uncle's condition and spent days, weeks, and months at a time trying to make him come back to himself. And so, from my earliest memories, I can hear my father saying to me that I didn't have to be perfect or smartest or best at anything. It was all right to be myself, to be as good as I could be. He would endlessly repeat that no matter what I did there would be someone better at it than I was—and that that was OK.

In both college and graduate school this bit of psychology was enormously helpful. I was never the smartest kid in class—even in math class. I went to Erasmus Hall High School with 8500 students (including Barbra Streisand and Bobby Fisher). I was 109th in my graduating class and damn proud of it. When I went to graduate school at the University of Wisconsin, Madison; there were over 300 graduate students in mathematics and it seemed like they were all better than I was. But I took my father's words to heart. I didn't need to be best. I needed to be good enough and more importantly as good as I could be.

The summer after my senior year I had an REU grant from the NSF. I used it to study from Kleene's, *What is Metamathematics?* It was a revelation. I had always enjoyed logic in college, if mostly from a philosophical perspective. But it quickly became clear to me that mathematical logic was the field I wanted to pursue. And since Kleene was at Madison, Madison is where I wanted to go. The timing was perfect. I graduated from college in 1963. The country was still in a post-Sputnik frame of mind. The large Midwestern universities had a ton of money for graduate work in math and science and they were extremely free with it. My parents were quite poor at the time. Brooklyn College, where I had done my undergraduate work, had cost me exactly $8 a semester tuition and fees. And I had a $350 a year New York State Scholarship to boot.

Madison offered me a teaching assistantship and free tuition. I was never so rich. With the princely sum of $2500 a year, living in a dorm, I sent $600 home to my parents. But I was on my own for the first time in my life. I had lived at home, a one-bedroom apartment, with my parents all through college. Now at the ripe old age of 20, I was living in a very different environment. And, like most of my fellow graduate students, I simply knew that four or five or six years from now I was expected to do some original research that would be considered significant enough to be given the club membership certificate—a Ph.D.

I should say early on that I loved doing mathematics, even when it was completely frustrating. There are few feelings that can be compared to the joy of discovery–when a proof falls into place and you realize that it could never have been any other way. Of course, that feeling is frequently followed by the sense that it was all obvious to begin with and a brighter mind would have seen it from the beginning, but the joy is very real. I was certainly helped by the work I had done with my dad. After all, I knew what it was to do original mathematics. Granted it was mathematics that hadn't been valued as important research for 100 to 150 years, but nonetheless it was original. To illustrate the point, in one of my father's papers, printed in the Monthly, he rediscovered the Napoleonic triangles. For a time he was quite crushed, learning that what he thought was totally new had been known before. But several of his results were new and deemed worthy of publication, so he got over his disappointment.

As for me, I was hard at work learning mathematical model theory at a time when the field was truly coming into its own. I did my graduate work from 1963-1967. These were golden years for logic. Paul Cohen did his independence proof. Michael Morley did his work on categoricity (probably still the most beautiful proof I have ever seen). It was the time of Tarski and Fefferman and my thesis advisor H. Jerome Keisler. The logic seminar was run by Steven Kleene and J. Barkely Rosser, Gödel's best Ph.D. students. This was a time of giants and I was privileged to walk among them. I was a sponge.

There was one sad lesson that I learned at this time. I would come home to New York on holiday and rush into my father's study to tell him what I was learning, especially the work in logic, abstract algebra, and finite geometries. But to my great surprise and disappointment, my dad was done. He made it clear to me that his mathematical plate was full with the definitions and theorems and problems in the domain in which he was working. He didn't have time and space for these new ideas. He talked about being too old to study these things, no matter how hard I pressed, no matter my enthusiasm. I came to realize after a while that I was his best student, but he would never be mine. I vowed then and there never to close myself off to learning new things, never to be too old to try, even if it was try and fail.

My doctoral research was on the undecidability of certain finite theories. As fate would have it, my love of geometry truly paid off. The main result in my thesis was a proof of the undecidability of the theory of finite projective planes. This problem had been open for quite some time. I was fortunate to see that a new method for embedding one theory into another, discovered by Michael Rabin and Dana Scott, could be applied to the finite planes. The proof was five pages. With discussion, history, and a few important lemmas, the thesis came in at thirty-four pages. But the main idea was there in an instant. One minute I'm a graduate student contemplating what I will do with my life with just a Masters Degree, and the next I'm applying for an instructorship at Cornell.

I remember calling my father when it became clear that my thesis would be accepted. He literally said that he could die a happy man. I wished him a much longer life. But the burden of fulfilling his life's dream was lifted. And when I looked inside myself, I realized that I didn't have anything truly of my own to take its place. What now? Well, I went to Cornell and began to teach in earnest and started to work on problems related to my thesis and the things I more or less knew how to do.

Blame it on the 1960s. I started to teach at Cornell in 1967. Things were happening all around me. The Viet Nam War protests were in full bloom. Daniel Berrigan was at Cornell during this time. Sloan Coffin would come up from Yale. A group of African-American students took over the student union building. Guns were smuggled in and the campus became extremely tense. I was 24 and hadn't thought about anything but mathematics until now. And it all came flooding in—the war, racial injustice, the counterculture—all of it.

I began to fully appreciate my strengths and limitations as a research mathematician. I knew a good problem when I saw one. I could create an interesting problem that others would see as important or at least useful. I was technically weaker than I should have been. I was a good collaborator. I worked and played well with others, so to speak. I was a talented expositor. And I learned that I was a natural-born teacher. Honestly.

It now became a question of values. First, a true story. On my first day at Cornell, Alex Rosenberg, who was the Chair at the time, invited me in for a chat. He told me that he hoped that I would be a good teacher and carefully prepare my lessons. But he also told me that, as far as my future at Cornell and the possibility of tenure was concerned, all that mattered was the quality and quantity of my research. He made it clear to me that, although this was not the way he thought the world should be, this was the way the world was. And, if I truly wanted to get tenure and had to make a choice between my teaching and my research, that research was the option I had to take.

As I said, I was 24, very idealistic, and hence was quite devastated. I knew that the gods must be crazy. I went home that night and wrote a letter to the then Secretary of Health, Education, and Welfare, John Gardner, told him the story, and asked for his advice. Much to my surprise, he wrote back. He enclosed a long reading list about U.S. higher education and the advice I had asked for, namely publish enough papers to get tenure at a respected institution and then, as a full-fledged member of the club, work from within to change the system. These are words that I took to heart.

And so, after three years at Cornell, I went to UCONN—the University of Connecticut—before either the men or the women were much good at basketball. It wasn't a happy eleven years. In fairness to the faculty there, they thought that they had hired a research mathematician and what they got was a budding mathematics educator. As it turned out, I had a good friend from my Cornell days who was a biochemist and was working at the Education Research Center at MIT. I went up to visit him one day and he introduced me to some of the people working on a mathematics education project. Interestingly, the staff of the project were all physicists under the direction of Jerrold Zacharias and Judah Schwartz. Basically, they were looking at ways to teach calculus using a variety of experiments that could be done by students at home.

I was hooked. I had never really seen a serious or useful application of mathematics save rates of change of ladders sliding down walls. This was the real thing—bubble accelerometers, wine making kits, microscopes with drops of water for lenses, etc. And so I began commuting once a week to volunteer my time and to learn. Of course, I brought many of the ideas back into my own calculus courses at UCONN, including some that were what was in pedagogical fashion at the time—mastery learning, group work, even portfolio assessment. For me it was all a brave new world. For the chair and some of the faculty at UCONN it was a lot less. To say that my efforts were neither understood nor accepted is an understatement. I had a knock down drag out fight over tenure. And, as I've said, for good reason. Fairly early into my teaching calculus through applications immersion, I pretty much gave up research in mathematical logic.

I will relate one incident as a mathematician that truly made me smile and I think would have given my father some "nachus" as we say in Yiddish. Two more logicians were hired soon after I came to UCONN and, as we were only an hour from New Haven, we began a weekly joint logic seminar with Yale. The first year it was held each of the logicians were given an hour to talk about some piece of our work. It was rather intimidating. After all, Abraham Robinson was the head of the Yale logic group, and a giant in the field. When my time came I think I told them everything I had done and pretty much everything I knew in that one hour. When it was over Robinson came up to me and said that my talk was very interesting, but he felt that I had gone a bit fast. I was in heaven for weeks afterward.

At any rate, by this time I was entirely focused on the teaching of mathematics through applications and modeling. And, as I learned more about curriculum development and proposal writing (NSF was then as now the major funder of these efforts), my self-image began to change. Before, I was a researcher and a teacher, now I was becoming a teacher and a math educator. I should say that I have always been lucky in my friends. Throughout the years I have met, known, and in some cases loved people of great moral and intellectual strength. Of course, as a research mathematician I had many role models. To a large extent that is what graduate school provides—people and an ethos to emulate. But I was extremely fortunate to find role models in mathematics education. Here I include Gail Young and Henry Pollak, both brilliant mathematicians and leading mathematics educators.

Getting to know and work with Gail and Henry was a little like having Abraham Robinson say something nice to me about my research. If these guys respected what I was doing, what did I care that the people at UCONN didn't? I was working with giants. And the giants liked me! One of the points I alluded to earlier was the extent to which taste plays a part in being a mathematician. You need the taste to pick problems that will be interesting to others (as well as important, hopefully, to the field). You need the taste and self-knowledge to pick a problem that you have the ability to solve or at least move toward a solution. And you need taste in your colleagues, picking people to work with or consult on problems for which you need help.

I've always had reasonably good taste in ideas. And I've had exquisite taste when it comes to recognizing talent and expertise in other people. I could not have achieved success as a mathematics educator without those people. And this brings me to the real turning point in my career—1981. In 1981 I took a sabbatical from UCONN and went to work at the Educational Development Center (EDC) in Newton, MA. EDC had inherited the Undergraduate Mathematics and its Applications (UMAP) grant from ERC which had been disbanded at MIT. UMAP was a project designed to produce modules in undergraduate mathematics that contained real and contemporary applications. I had a part in writing the original proposal for this grant in 1976. And, embedded in that proposal, was the idea that an organization would be formed to carry on the work after NSF funding ended. After all, we argued, there was no fixed set of applications of mathematics. It wasn't as if we would "finish" writing modules about them all. New applications were being discovered every day and we wanted a vehicle for getting those applications into the classroom. At any rate, the idea of COMAP and even money for incorporation as a not-for-profit company were embedded in the UMAP grant. And, when I went up to Boston, I decided to stay. I wound up taking three consecutive leaves without pay and finally, after four years of living on my own wits, gave up tenure to work full-time on mathematics education as my own boss.

At the beginning of this article I said that I would attempt to contrast the psychology of being a mathematician with that of being a mathematics educator. I realize that, in order to explain the mathematics educator part, I need to break it up into two pieces. One of course is the intellectual piece. The other deals with the nature of the beast, the setting in which the work takes place. Most mathematical research lives in a university environment. Yes, some of that work may be grant supported, but unlike the laboratory sciences we can work with very little stuff. One remembers the old joke about mathematicians needing some paper, a pencil and a wastebasket–and sometimes not even the wastebasket.

But my work in mathematics education has always taken place outside of academic institutions and most of it has been very dependent upon soft money. Getting that money has been a large and important part of what I do—and the psychology of writing grants and pleasing funders deserves its own description. I should say that, almost from the very beginning, there were a number of colleges and universities that offered to house COMAP. The decision to be independent of any particular institution was primarily based on not wanting our materials to be seen as targeted to any one audience. We feared that, if we were at College X, then we would be seen as producing material for College X's students. And I must also admit that living through the academic politics at UCONN had left a bad taste in my mouth. I simply wanted to succeed or fail on my own.

Just one last family aside. My father was devastated. He had had a dream for me; he had seen me fulfill it; he simply couldn't understand why I would give up a tenured university position for the world of business. For my father, the possibility of another Depression was never far from his thoughts. Teaching, tenure, these were protections—protections that he wished he had had. Being a university mathematics faculty member was his highest aspiration—one I had achieved. Giving it up, in his mind, was the height of folly if not mental illness. I admit that there were times over the past thirty years where I saw his point.

OK. The psychology of a mathematics educator. First and foremost, research in mathematics and development (I am after all a curriculum developer at heart) in education live with different time scales. When one does mathematics there is a sense of timelessness. If the problem is interesting to the mathematician and someone will publish the result, it really isn't crucial that the ideas become popular or that the theorems have immediate application. Who knows when an idea may be seen as important? Who can tell when a piece of mathematics will find application? That's not the job of "pure" research. Truth for truth's sake.

But in education at heart we are trying to help children (or adults) learn. Where the mathematician's articles of faith revolve around the pursuit of truth, intellectual honesty, and new mathematics, the articles of faith of the educator are of a different nature. There is in educational work a fairly large dose of social justice. We want all kids to learn to the best of their abilities. We honestly hope to leave no child behind. Those of us who love mathematics and have seen it enrich our lives want it to enrich the lives of others. If mathematics is so central to modern society and daily life then we want others to recognize and understand that centrality. Teachers are mathematics ambassadors and as mathematics educators our job is to give them all the help we can so that they can do their job. And time is our enemy.

Creating good curricula or assessments using print or video or software, doing the necessary testing and revision, all takes time. But kids are failing now. New teachers are coming out of education programs unable to explain to students why they should learn mathematics. Experienced teachers are overburdened by a failing infrastructure, and social injustice, and bad nutrition; and a host of real real world problems face too many children every day. That is the environment that the mathematics educator works in, and there is never enough time.

Time plays another role here. When the work is undertaken with grant support, the funding agency wants to see progress (often measured by improved student test scores) within the life of the grant or within the term of office of the politician who initiated the program. Remember America 2000? We were to be the best in the world in math and science by then. Actually I believe that there is still a law on the books that says we will be metric by 1975. So the work takes time. Funders don't have the needed patience. And generation after generation of students are ill-served. Oh, and every twenty years or so, we reinvent the wheel.

But back to the psychology of the mathematics educator or in my case the curriculum developer and grant writer. I have articles of faith. These are things I believe and are part of my work because in my heart I know they're right. I don't test them or read studies by people who do. I don't care. These are the tenets of the religion of my vocation. I believe that it is a positively good thing to teach people how mathematics is used. I believe that mathematical modeling is a life skill and should be taught for its own reasons apart from the reasons we teach mathematics. I do not believe that all students can learn mathematics equally, but I believe that all students can (and should) learn mathematics. I believe that students learn differently and one size does not fit all. And I believe that the curriculum should be a mile-wide and an inch deep, because there is so much beautiful and new mathematics and applications of mathematics that students should be exposed to. And I couldn't care less about AP calculus and international comparisons, especially those that test curricula skills.

I don't know if I always believed these things. Actually I can't honestly say when this worldview came together. But I can say one thing for certain—it never would have come together without Gail Young and Henry Pollak and most importantly Joe Malkevitch. Joe and I were in graduate school together in Madison. We were floor-mates our first year there. We lost touch with one another after graduation. Then, as luck would have it, we bumped into each other at the joint mathematics meeting in St. Louis in 1979. I started to talk with Joe about my ideas for COMAP and we struck a common chord. I'm not a history buff and can't really talk knowledgeably about the relationship between Lenin and Trotsky, but Joe became the philosopher and theoretician for essentially every project from UMAP until today. Sometimes I forcefully stop myself from phoning Joe, because our conversations never last less than an hour and a half.

So I honestly don't know where his ideas begin and mine end. Joe taught at York College of CUNY for his entire career and, though his ideas run throughout the proposals COMAP has written, it was my job to bring the various teams together, write the grants, shmooze (a technical math education term) the funders and bring in the money necessary to do the work. A word or two needs to be said about writing proposals. This is an unnatural act. You find yourself using words like "metacognitive" and "facilitate" and citing work you don't respect (because the funding agency does). Uri Treisman and I used to play a game where we would give ourselves points if we managed to insert certain words in our proposals. As

I recall "walrus" was worth twenty-five. The highest number of points, fifty, was to go for "Troy Donahue." Neither of us ever succeeded in getting those points.

I have often joked that I hoped to found the Journal of Unfunded Proposals, because that's where you will find some of my best ideas. And that really is not a joke. Of course proposals have to fit within the guidelines of the program one is applying to but, given the current panel system it's a little like approval voting. Basically, you are rewarded for being somewhat middle of the road and penalized for thinking too far out of the box. One extremely negative review can pretty much sink a proposal these days. I wish I had a proverbial buck for every proposal that was turned down with some variation on, "This is a good idea, with good people, but it's too innovative/different to work." In the days of the "old boys club," my job was pretty much to convince one program officer. Today it doesn't pay to be flamboyant.

Proposal writing is almost the exact opposite of doing research. When one is doing research, the problems and the ideas seem terribly important. When the breakthrough is made there is a moment of euphoria. But that is almost immediately followed by a sense that the work was trivial; that a better mind would have seen it immediately. With proposal writing, you feel like a prostitute, writing things you know that the funder wants to hear, couching what you write in the language and jargon of the day. And of course, getting it out the door just in time to meet the funder's dateline. But once it's done and you know that the proposal is under review, you find yourself thinking that no sentient being could turn down such deep ideas so cogently presented. And remember, in most cases, six to nine months go by from the time a proposal is submitted to the time you hear. And all during that time you are fielding calls from members of your team asking whether they should make plans for other work during the summer or next semester.

I want to be clear. There is a great deal of creativity in honing an idea, in choosing the people to work on it, and even in finding ways to make expressive arguments about why the idea deserves funding. But the joy is in the doing. When you get the call that the project will go forward (remember those nine months) job one is to call the team together and ask yourselves what this project should really be about. This is an important truth—projects and proposals are not the same thing. Proposals are not contracts. They are a set of ideas. And with multi-year programs those ideas will change—should change. As you work on a project you learn; and as you learn you have to adapt. Funding agencies understand this, which is why they ask for interim reports to track how the project is progressing and what new directions it might take.

But all of this is an abstraction. The projects that I have worked on over the years have made things—modules, books, TV series, modeling contests. These are all very tangible things most of which you can hold in your hand. There is nothing quite like the experience of leafing through a book that you have worked on for years when it comes fresh off the press. The smell is as joyful and distinctive as freshly baked bread. There is a sense of accomplishment that is truly different from the moment of discovery of mathematical research. It feels enduring. Of course, not all books sell; not all books stay in print; not all books are used by students across the country. But, nonetheless, knowing all that, when you see that book in your hand, it feels enduring. I know that this is all changing as the world goes digital. And I'm sure that it will save a lot of trees. But I for one am glad that I lived in a time when I could make books of paper.

There's another psychological point to be made. I think of myself as a revolutionary. COMAP was a revolutionary idea. The Mathematical Contest in Modeling (MCM), the *UMAP Journal*, *For All Practical Purposes*—all revolutionary ideas. But I just read proofs of volume 35.2 of the journal and MCM has been going for 30 years and FAPP is going into its 10th edition. I am smack dab in the middle of the establishment. Moreover, the ideas aren't revolutionary any more. When the *UMAP Journal* was founded, it was the only place that articles about teaching mathematics through applications could be published. Now we have to fight to get enough articles each issue because so many journals are available to publish them. The new Common Core standards feature mathematical modeling as an essential practice, whether the authors really meant it or not. COMAP was formed primarily because work in applications and modeling, especially curricular work, needed a home. Modeling and applications need not be homeless today.

Don't misunderstand. I have no intention of closing shop. COMAP still brings a unique perspective. While the overall mission of enhancing the place of applied mathematics and modeling in our nation's schools and campuses has remained the same, we have moved closer and closer in the classroom to the implementation of curricular change. That is a lesson I learned kicking and screaming. I love to make good things. The line from the Tom Lehrer song comes to mind, "I just send them up. Who cares where they come down. That's not my department says Werner Von Braun." But it is my department. No one has taught me that lesson more clearly than the current chair of the COMAP board of trustees, Eric Robinson.

This is different from mathematical research. This is not about proving a beautiful theorem or making a beautiful book. This is about schools and teachers and kids. If we don't change their lives then we are simply playing at being educators. Of course there are mathematics educators who sit in academic departments and write papers and go to professional meetings and look indistinguishable from research mathematicians. But we cannot afford to separate research from practice in education. I believe this as strongly as I believe anything. Yes, it's encouraging that mathematics education has become a discipline respected in its own right. But those who work in the field must know both mathematics and education. Knowing mathematics means knowing what it is to do mathematics as well as a broad understanding of the subject and its applications. Knowing education means knowing the classroom from the inside.

For better or worse, I think that in the future there will be fewer Sol Garfunkels and Gail Youngs and Henry Pollaks and Joe Malkevitches and Eric Robinsons, trained as research mathematicians but devoting themselves to work in mathematics education. What is important is that the graduate programs that train people to do research in mathematics and the ones that train people to do work in mathematics education can find ways to work together. For the truth is, as my career illustrates, one can do and find joy in both.

16

Dynamics of Mathematical Groups

Jane Hawkins

What is a mathematician? Not who, but what. Are there axioms or is there a lemma whose hypotheses, if satisfied, allow you to conclude "then you are a mathematician"? Clearly not. Regardless there are some conjectural characteristics that come to mind. Most of us majored in math, and in doing so decided before the age of 20 that we wanted to study math. At some point most of us studied math to the exclusion of all but closely allied subjects. However this is a fairly modern definition; the history of mathematics and mathematicians is much sloppier than that and filled with fortuitous events that led to mathematical discoveries we take for granted today. Nicole Oresme was a Catholic bishop in France in the 1300s during a time when astrologers were members of the intelligentsia; he also held the title of Secretary of the King and wrote popular articles about science by translating Aristotle's work into French. At that time courts often kept on hand astrologers to help predict significant events, but Oresme was strongly against the acceptance of astrology as a science. Philosopher, priest, mathematician and all-around scientist, Oresme used probabilistic arguments in his analysis of nature, long before probability existed. This led to the following admission that he wrote more than once: "... (except for the knowledge of true faith) .... I indeed know nothing except that I know that I know nothing." He also came up with the earliest notion of graphs appearing in *Tractatus de latitudinibus formarum* and shown in Figure 1, which plot velocity against time of a moving object. Three hundred years later, Pascal gave the formal mathematical underpinnings of probability theory in order to solve a problem of how to calculate odds correctly when a gambling game is interrupted.

So early mathematicians were not focussing exclusively on mathematics. Oresme's motivation for working out properties of irrational rotations on the circle is well-documented, and brings to mind this hummable song from the late 1960s musical *Hair*:

> When the moon is in the Seventh House
> And Jupiter aligns with Mars
> Then peace will guide the planets
> And love will steer the stars

Dynamics of Mathematical Groups

Figure 1. Oresme's graphs of velocity vs. time from the 1300s.

Indeed, Oresme developed the notion of incommensurability of rotations of the circle, irrational rotations, and specifically discussed their aperiodic orbits in order to debunk the ideas of astrology. In other words, to prove that Jupiter will never again align with Mars while the moon is in the seventh house, my favorite research area of ergodic theory was born. This was no fluke; Oresme knew math and physics; while he predicted that it was quite "probable" that the earth moved, it would be another 200 years before Copernicus proved that the sun, not the earth, was the center of the universe.

These are some origins of the mathematics I work with daily, and why it is important that we not try to define what it means to be a mathematician. Nevertheless the purpose of this essay is to give some handwaving arguments and proof-by-example descriptions of what makes mathematicians tick. Translating the mathematical parlance of the previous sentence it means I claim little, will prove nothing, but hope to explain a few of our professional tics.

Mathematicians from the history books conjure up images of men, and rightfully so. Most of them were men. By now we all have heard a few of the depressing stories of women mathematicians before and in the twentieth century. Hypatia, daughter of a well-known Greek mathematician, was a respected philosopher and mathematician. She was accused of being a pagan enchantress and was stripped and torn apart limb from limb by a mob of Christian zealots in the streets of Alexandria in 415 AD. She is alleged to have said: "Reserve your right to think, for even to think wrongly is better than not to think at all," (but since many believe that none of her writings survived, this quotation may not be accurate).

Maria Agnesi, born in 1718 in Milan, analysed an important planar curve which due to a so-called translation error earned her and the curve the nickname Witch of Agnesi.

In 18th century France, Sophie Germain sent comments to her math professors, sometimes including original proofs, using a male pseudonym,"M. le Blanc," so they would be read.

Ranking eighth on the Tripos exam at Cambridge in 1880, Charlotte Scott's name was left off the list of top graduates that year and no Cambridge degree was awarded to her because she was female.

Some of the stories have cheerful details. Mary Everest Boole, married to George Boole, was a self-taught mathematician and a mother of four daughters. She wrote several books that were attempts to make mathematics more accessible to children. Some of them changed approaches to education completely.

From her 1909 book *Philosophy and Fun of Algebra*, she makes this slightly offbeat plug for learning algebra.

> Many people think that it is impossible to make Algebra about anything except number. This is a complete mistake. We make an Algebra whenever we arrange facts that we know round a centre which is a statement of what it is that we want to know and do not know; and then proceed to deal logically with all the statements, including the statement of our own ignorance.

She is famous for teaching children math through sewing—in fact they were making string envelopes of curves. Rumor has it that Mary co-wrote a lot of George's papers.

I prefer stories of mathematicians' intellectual valor. I will relate one of my favorites below, but we should not ignore the humbling accounts of women in math. The anecdotes of discrimination against women in math are still legion; who among us hasn't witnessed a few or experienced many? I could tell you stories that could shock you but you might not believe them, and in any case, that's a different essay.

Having broached the subject, here is a sample of what has been said to me.

(1) We don't think you're serious about math so we want you to apply for graduate schools in Europe so you're not competing with the men here. (I received my Ph.D. from the University of Warwick on a Marshall Scholarship.)
(2) Women always sleep with their male coauthors.
(3) Does your having a child mean you're never going to do research again?
(4) Women shouldn't be allowed to attend conferences with men. Too distracting.
(5) I'd like to have you as my Ph.D. advisor but I don't think women should be working outside the home; my wife stays home.

There follows a sample of what else I've heard said, some of it fairly recently.

(1) She is much too good looking to be doing math. She should find a better profession.
(2) I don't think I have ever seen a blond woman doing math before. Are you serious about math?
(3) There is no way I will even consider hiring someone's girlfriend no matter how good she is.
(4) What if we hire them both, they break up, and we're stuck with her?
(5) What area of math do you work on related to this conference, as if I care?

Interspersed with comments of this sort has been a thread, an unbreakable one, of encouragement and support and this is probably true for most mathematicians and probably for all the women whose research continues beyond their Ph.D.s. There was a male math teacher in each of my elementary, middle and high schools who singled me out for my math ability. In third grade, before I knew that life has hardships in it, my teacher announced to the class one day "I think writing down answers to our weekly Times Table Quiz is hampering the progress of some students." The goal was to know single digit multiplication so well that we could complete timed random worksheets in five minutes or less. He then asked me to stand up and call out the answers, and showed how my best time was cut in half from three minutes to one and a half minutes without a pencil. He asked a few others to do likewise, and they fumbled around a bit and did poorly, and then the exercise was quickly ended. I shared a moment with him at that instant when we realized I was different. My grandmother, a second grade teacher, bought me a special Life Science Library book called *Mathematics*, published in 1963 before I turned 10, that I pored through on my visits to her house. The book is a treasure trove of ideas from calculus, probability, and the newly emerged field of topology, written for the layman; I was completely intrigued by the ideas presented.

Even so, I did not take calculus in high school and declared math as my undergraduate major only as a back-up plan since I knew I was good at it. I was looking for "a better fit" for something to study; I never found one. Mathematics really started to come to life as a potential career in graduate school at Warwick. I had never met people my age who sat around discussing math for fun instead of to prep for an exam—and professionals came from all over the world to sit in the Common Room sipping coffee and tea for the same purpose. The late 1970s was a very exciting time to be at Warwick and working in ergodic theory and dynamics; however the good mentoring and advice I received there came from many directions. Ian Stewart, Dusa McDuff, David Epstein, and Christopher Zeeman all advised me, and most importantly the advisor I ended up working with, Klaus Schmidt, was highly encouraging and came as part of a robust and enthusiastic dynamical systems and ergodic theory package at Warwick. Even at the time I felt fortunate to have landed at Warwick so the path there mattered little to me then.

Considering that educating undergraduates is what gets us our paychecks, most mathematicians have an oscillating view of teaching. Some days student interaction is the highlight of a long week with nary an original idea. Other times it is taxing to go into a room and compete with rows of laptops, smartphones, and iPads, while a research paper lies bubbling with energy but untouched. Professional attention to undergraduate instruction has gone in and out of fashion during my years in the field. There is a consensus agreement that we deliver instruction in mathematics so students pick up on the excitement of the field and want to learn more, and to prepare them for a profession in which good analytic mathematics skills will translate quickly into successful work skills. But there are as many ways to achieve that goal, or fail at it, as there are practitioners of the trade.

My own introduction to teaching was a baptism by fire. Since I did all my graduate work in England, I had little idea that in the United States graduate students had teaching duties and hence training. Moreover, as a Marshall scholar, apart from auditing some undergraduate classes, I had little contact with undergraduates. The classes I sat in on were courses in measure theory and functional analysis, taught on the level and in the manner that we see in the first few years of American graduate school, but taken only by advanced British third year undergraduates. There was no dialogue with students and very little eye contact; the professor went through the syllabus on the board, pausing briefly to take a breath, and off he or she went again. (I took three or four courses from women faculty.) There were no student teaching evaluations and no pauses during class to see how we were doing out there. We were taking notes, listening, and sometimes thinking; that's what we were doing. If you thought the professor made an error, you did not interrupt; it was typical to meet with some classmates and your notes after class, go through the material carefully, find errors and determine if they were yours or the prof's. If there was a genuine error then one or more students would carefully approach the instructor at tea or during office hours. Should I be ashamed to admit that I thought the system worked? The daily tea attended by virtually all faculty and graduate students was a key feature, but I thought it worked.

Upon my arrival at Stony Brook for my first job, I was assigned to teach two courses: Galois theory, which I had never taken, and probability theory which had 125 enrolled students and met in an amphitheater. I was 25, lefthanded (this related to my messy use of chalk), too short for the boards in the large classrooms, and simply a terrible instructor. I followed the only model I knew: entered the room and nodded to the class, turned my back to the room, and proceeded to reel off the material from my notes, much of which I had learned the week before. I had terrible stage fright, and also since most of my math learning took place in England, stumbled over "zee" and "zed," the pronunciation of Greek letters, and in combinatorics problems attracted snickers by saying anticlockwise instead of counterclockwise. It wasn't long before my undergrads asked the Stony Brook grad students what was wrong with me, thinking I must be a student. Then the advice came trickling in. Most of it was quite helpful too—stop regularly and look at the students, see if they have questions, explain what you're about to do and/or what you've just done. Before each class my stomach would clench up and I would start perspiring profusely; to say that I had no sense of humor while teaching is an understatement. One day, one of the graduate students who was dating a math major in my class came to offer more specialized advice. By this time my teaching abilities were legendary among the math majors and graduate students—Writes Fast and Never Turns Around. He said "You need to have more fun in the class. You should actually say funny things once in a while; make the class laugh. You have a sense of humor outside of class. One of my friends wears a gorilla suit from time to time to teach in. You should think about that." That definitely was funny, thinking of myself sweating in a gorilla suit with my back to the class, copying field extension properties from my notebook to the board. Still, I took his point and relaxed a bit and started making eye contact with the class even though I'm quite sure he meant the suggestion literally.

After two years of painful learning by doing, I was a halfway decent teacher, at least that's what the Undergraduate Director thought. My early teaching disasters had gone unnoticed by the faculty except for the visiting prof from Rice whose office was next door to mine. (He offered a sympathetic ear and we got married a few years later.) The Undergrad Director called me into his office and said "I hear you're a good teacher. That's good, but not so good. Didn't anyone tell you we fire our good teachers?" It was said half in jest, but after checking around I found out that it was generally accepted that if your teaching was good enough to attract compliments, then you weren't spending enough time on research. Given the outcome of that job, all modesty aside, my teaching wasn't really all that good. My years at Stony Brook helped form my current worldview on getting tenure in a research department; I've visited and taught in many departments since then and been in my current job for decades. Instead of giving you the thousand words, I refer you to my Faculty Performance Evaluation Charts in Figure 2.

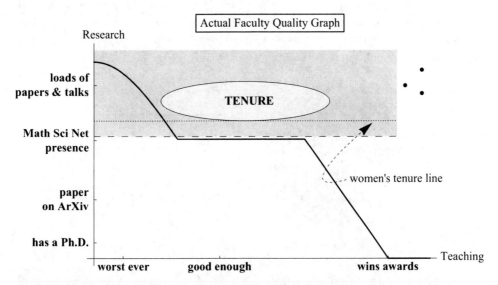

Figure 2. How faculty performance is evaluated.

Most of us do not have deep mathematical knowledge by the time we start our first job. Becoming a competent research mathematician involves a long apprenticeship working with the best people in our field for as long as we can until we are ready to launch our own research program. What math we know and how we learned it has changed quite a bit in the 3+ decades I have been studying math, and it is divided a bit by generations. Philosopher Marshall McLuhan was prescient in his 1960s book about the impact of media on our culture. At the time he wrote *The Medium is the Massage*, a play on his phrase "the medium is the message," the first minicomputer had been invented. While it is likely that he was referring to television more than computing, now that the line between TV and computer has been blurred it doesn't matter; he was right. There is no doubt but that computers have transformed all areas of pure and applied mathematics. Probably each

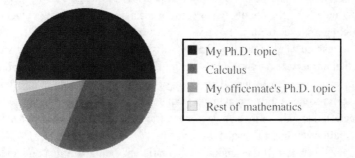

Figure 3. Mastery of mathematics by new Ph.D.s in the 20th century.

of us has some relationship to at least one computer, and by relationship I mean that if the device isn't working or available we suffer withdrawal symptoms. But what does a computer do for a mathematician? The answer varies: sometimes it retrieves math papers others have written, sometimes we use it as a desktop publisher, and a computer does a lot of fast calculating, estimating, and visualizing. Let us not overlook the symbolic and logical manipulation too. For me, computers suggest theorems to prove. I love to iterate mappings on my laptop to obtain output in visual form. From there I can guess what is probably going on mathematically and start to prove it.

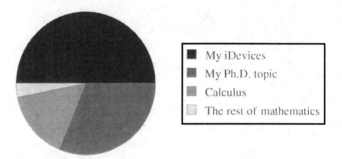

Figure 4. Mastery of mathematics by new Ph.D.s in the 21st century.

Choosing a good math problem to think about and solve is simultaneously very personal and often a byproduct of a chance encounter with another mathematician at a conference, a colloquium, or even in an airport. A propos of working on a good problem, I am a fan of Henri Lebesgue. It takes real guts to write a Ph.D. thesis where you invent a new type of integration. There's an apochryphal story about his Ph.D. defense in Nancy, France in the early 1900s. Henri is standing at the blackboard being barraged by questions about his newly defined integral. He is sweating profusely and one of his questioners asks: "But why do you want a new integral? What is wrong with the Riemann integral?" Lebesgue pulls his handkerchief from his pocket and wipes his sweating brow. He crumples it up to return it to his pocket and pauses. He holds up the crumpled cloth and says "So we can integrate functions with graphs like this." Okay, surely you are thinking that unless he tore his handkerchief, it still formed a continuous if not differentiable surface, so the Riemann integral would have been just fine. Nevertheless his sweaty kerchief represented a nasty function, so I like the story.

I was on a panel with women mathematicians a few years ago and a young successful panelist had this advice for the audience composed mainly of graduate students and new faculty, all female. Never be afraid to try the hard problems. Hard problems that interest you a lot are ones you can likely solve. To embellish what she said a bit, there are problems in any field that attract a lot of time and attention. They resonate on some level due to their simplicity, their demand for technical prowess, or their duality to something you know. Go ahead and look at some of those; maybe you can solve one or maybe you can get in on some interesting action that will keep you busy for a while. You could end up with a mathematical object named after you like Lebesgue did.

There is no shortage of jokes about mathematicians being weirdos and social misfits, with punch lines about whose shoes the guy is staring at (he/she stares at your shoes if an extrovert, and at his/her own shoes if an introvert). In my experience we're not as strange as all that. In fact, most mathematicians are quite well travelled, well read or musically inclined, and while not exactly well dressed, have strong opinions about what they wear. I have dined with colleagues at restaurants sought out for their outstanding pierogi, bulgogi, or barramundi, and witnessed sommeliers being quizzed before a bottle of wine could be uncorked at the table. I have walked miles to a pub in England for the perfect and only acceptable ale. While I'm too squeamish to eat any raw fish with the property that an error in preparation asphyxiates the diner, I've been present for many a conversation where the virtues of fugu are extolled. From what I have seen if a mathematician is staring at another's feet it is to determine if the soles of their shoes give better support than the ones being worn, in which case they will change brands. So why do people insist that we are strange? Because as a group we are extreme in our preferences.

I have noticed that many mathematicians, myself included, have an algorithm for dressing that isn't used by the general population. I mean the average person just doesn't use an algorithm for dressing, but hear me out on this. It makes sense to establish a dress code that minimizes the amount of time deciding what to wear while maximizing the ability to teach, give a public talk, and walk comfortably. So we each have a "uniform," and the men are just as particular as the women on this point. Uniforms run along the lines of: black pants, polo shirt, Nikes; or jeans, plaid shirt, New Balance. And it is often the case that the mathematician has a closet full of them. I set up my own wardrobe at the beginning of the semester: line up cotton t-shirts, pants, cardigans, clogs. Choose one from each category each day; pretty simple.

At a recent AMS (American Mathematical Society) governance meeting I got into a conversation with a few women mathematicians I hardly know, all of us dressed "one level up" for the administrative occasion, and we discovered we were all wearing the same brand of clothes: clothes with simple lines and colors that are easy to wash and wear, so worth their high price. I was surprised that I adhere to the stereotype, but why not? Saying that mathematicians rarely change their clothes is an oversimplification; if black pants match everything, why buy red ones?

Another way in which we mathematicians are often perceived to be antisocial is in our everyday language. By training we use intense focus to produce and solidify ideas that can be communicated only through precise language. Woe unto anyone using superlatives in our presence, family and friends included. If someone says "This is the best movie ever made" to a mathematician, it opens the door to hearing lists of other films considered by the arguer and critics, dating back to before talkies, Oscar winners and nominees, all that might possibly be better. An alternative response might be simply "I don't care about films" and the subject is changed. It's an amazingly unpopular trait at family gatherings; I see how it gets us a bad reputation. Putting this tendency towards literal interpretation to good use, at a math conference I sometimes make a statement along the lines of "Don't you think that was the best talk on xxx ever given?" It's my way of flushing out information on all talks given in the past forty years on that topic, because that will always come out immediately and in earnest with no talk left undiscussed.

Many mathematicians are addicted to travel; some will accept every invitation to speak anywhere, and funds permitting, every invitation to attend everything. It is because of this that we have been able to develop our views on food, music, and dress. However this

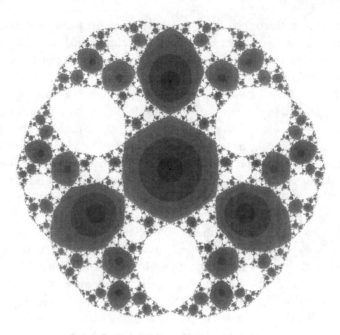

Figure 5. A Julia set in $\mathbb{C}$ with symmetries.

personality trait can wreak havoc with home life and also takes its toll on your health. Deep vein thrombosis is not terribly uncommon among research mathematicians age around 60 or up; nor is divorce. A few years ago, returning from a large overseas conference, I was on the same flight as a conference colleague. He had been diagnosed with deep vein thrombosis so every hour or so I could see him walking around the plane, a wild-haired bearded man bobbing up and down above the seats, doing deep knee bends in the middle of the aisle. Even though the airline tells us it's the proper thing to do, most don't, so it attracted more attention than the food carts. We then changed planes stateside to head to our respective homes and once again were in the terminal together. He asked me if I would keep an eye on his backpack, then lay down on his back on the floor, and shutting his eyes he put his feet up on a bench where travellers sit. This made a bit of a scene in our security cautious times. I sat a few seats away reading a great preprint of his, and at some point an airport official came over, walked around his large frame taking up a significant piece of floor space, and then stood near him. The uniformed worker then wondered aloud if anyone knew him or what was wrong with him. Even though I didn't know him well, I jumped up to defend his behavior, explained he was fine as far as I could tell, and was sleeping off the overseas trip while taking care of his circulatory problem. The uniforms starting to gather dispersed, my mathematical friend simply opened one eye and said "Thanks." After a bit of non-conversation we headed off to our respective gates.

There are many ways in which I think mathematicians are cool and wonderful people, a primary reason being that they love math as much as I do and will talk about it at almost any occasion. I've been on some great hikes with math friends, I have had some fascinating conversations over good beer. It is quite liberating to discuss math while hiking since you can't write anything on paper so you have to keep all the ideas afloat in your head. Over

Figure 6. The same Julia set on the sphere.

good beer math is great, but sometimes it seems like my head is afloat more than my ideas. I don't have deep vein thrombosis (yet) and I'm still married to the same guy I met at Stony Brook thirty-two years ago, so I don't think I'm addicted to travel. However I can pack a carry-on bag for two weeks in about an hour so definitely show those tendencies.

You might be a mathematician if your doodling is mathematical, you wake up and think about your problem before getting out of bed sometimes, and your down time is spent

Figure 7. Here is that same Julia set, well-defined on a Steiner surface.

working on smaller or back burner math problems. If one computer is perpetually running programs that help suggest theorems, draw pictures, plot things, or compute, you may be one of us. If in addition you have pads of paper lying around with math on them, and if a delayed flight means maybe another step in a proof can be understood, you're probably there. Some of my recent computer doodlings are inspired by the connections between complex dynamics and algebra talked about by Joe Silverman. I am now looking at Julia sets with a lot of symmetry that give well-defined Julia sets on the real projective plane as realized for example by a Steiner surface or Boys surface. I am working with topologist Sue Goodman and we have consulted `Mathematica` guru Mark McClure to get the pictures. Now it's a question of correct statements of the theorems; but these are now jumping up and down waiting to be written down.

The blending of analysis, topology, algebra, and computer algorithms keeps me amused for days on end—actually decades. Returning to the question that I posed in my opening sentence, I have no answer. And I need to get back to my math.

17

Mathematics, Art, Civilization

Yuri I. Manin

To the memory of Friedrich Hirzebruch.

Imagine Paris in the 1920s, the capital of modernism and high fashion. Reminiscing about those times, Coco Chanel told Paul Morand about Picasso:

> It was his painting I admired, although I did not understand it. I found it convincing, and this is what I like. For me, it's a logarithm table.

Think about this remarkable parallel. Mathematics is abstract, and the art of Picasso is abstract. It would seem that this is the most evident resemblance between the two imponderables: "Harlequin with Violin ('Si tu veux')" (1918) and a Table of Logarithms. Chanel, however, chooses a different word: both are convincing, and that is what attracted her.

* * *

In this essay, which is devoted to various aspects of mathematics as a long-term human activity, I want to pay special attention to this quality of being "convincing."

On a personal level, whether or not a proof, an idea, or a computer simulation is convincing depends on the mathematician's predisposition toward geometrical or logical reasoning, philosophy (conscious or subconscious), and a system of values.

On a social plane, large-scale historical circumstances come into play; they can cause either an amazing flourishing of mathematics, or its virtual extinction.

For obvious reasons, historians of mathematics study the places and times where mathematics was created or at least accepted as a legacy of other times and/or societies. But it would be very interesting to take a good look at the historical circumstances when mathematics was neglected or even (temporarily) disappeared.

The development of ancient (mainly Greek) mathematics in Europe was halted for at least the first thousand years of Christianity. Still earlier, before the advent of Christianity, when the practical and militaristic Romans created their higher culture, they incorporated the Greek humanities but not Greek science. Even the obvious military applications did not tempt them. According to Plutarch, at the siege of Syracuse the Roman general Marcellus vainly urged his soldiers not to retreat:

> What! Must we give up fighting with this geometrical Briareus, who plays pitch and toss with our ships, and, with multitude of darts which he showers at a single moment upon us, really outdoes the hundred-handed giants of mythology?

According to an interpretative tradition, Marcellus did intend to convert Archimedes to the service of Roman military (as Werner von Braun was converted later), but the sword of an ignorant soldier killed the geometrical Briareus.

According to the same tradition, Archimedes himself did not regard his engineering endeavors as "applications" of his mathematics—for his great mind they were just an unwanted distraction from mathematics.

* * *

The meager mathematical heritage of ancient Rome includes the number system

$$I, II, III, IV, V, VI, VII, VIII, IX, X,$$
$$XI, \ldots, L, \ldots, C, \ldots, D, \ldots, M,$$

which is still used today, mostly for ornamental purposes. The most instructive way to view it is as a unique archaeological relic of an archaic state of mathematical thought.

The unit "I" symbolizes a notch on a stick (and not the letter I, which was a later reading). Because of the effort needed to make each notch and the amount of space it required on the stick, they had to abandon the dumb but completely systematic and potentially infinite number system

$$I, II, III, IIII, IIIII, IIIIII, \ldots$$

and adopt a much less logical system of "names" rather than symbols

$$I = 1, \ V = 5, \ X = 10, \ L = 50, \ C = 100, \ D = 500, \ M = 1000.$$

Even though it does not permit one to go to infinity, for small numbers this system is cozy and efficient. Short sequences of these symbols are interpreted using either addition or subtraction:

$$2009 = MMIX = M + M - I + X.$$

Of course, there is no name for zero. The horror of "absence," "emptiness," is deeply ingrained in human psychology. Already in Ecclesiastes we read:

> ... that which is wanting cannot be numbered ...

The absence of a symbol for zero greatly hindered the development of the Roman system and its transformation into a positional number system.

* * *

The dissemination of a positional number system in Europe after the appearance of Leonardo Fibonacci's *Liber Abaci* (1202) was, in essence, the beginning of the expansion of a universal, truly global language.

Its final victory took quite some time.

The book by Gregorio Reisch, *Margarita Philosophica*, was published in Strasbourg in 1504.

One engraving in this book shows a female figure symbolizing Arithmetic. She contemplates two men, sitting at two different tables, an *abacist* and an *algorist*.

The abacist is bent over his *abacus*. This primitive calculating device survived until the days of my youth: every cashier in any shop in Russia, having accepted a payment, would start calculating change clicking movable balls of her abacus, "счёты".

The algorist is computing something, writing Hindu-Arabic numerals on his desk. The words "algorist" and modern "algorithm" are derived from the name of the great Al Khwarezmi (born in Khorezm c. 780).

The symbolism of this engraving was very much alive when the Reisch book appeared.

The Catholic Church supported the Roman tradition, usage of Roman numerals. They were fairly useless for practical commercial bookkeeping, calender computations such as dates of Easter and other moveable feasts, etc. Here the abacus was a great help.

The competing tribe of algorists were able to compute things by writing strange signs on paper or sand, and their art was associated with dangerous, magical, secret Muslim knowledge. Al Khwarezmi's teaching became their (and our) legacy.

* * *

The positional decimal notation, when it won, developed into the unique universal language of modernity.

The *semantics* of this language was the ability to count *absolutely anything*—notches, cattle, ships, florins.... Its *syntax* was determined by a general rule for translating an abstract quantity to (decimal) positional notation, and conversely.

Finally, the *pragmatics* of the language had two sides.

When the numerical text referred to a real-world quantity, such as in trade, higher level syntactic rules provided a key link between the text and the physical world. A famous example of these rules is the double-entry bookkeeping system that was codified by Luca Pacioli in 1494.

On the other hand, when the numerical text referred to scientific data, for example, astronomical observations, its pragmatics could be connected with the prediction of an eclipse or the construction of a quantitative model of the solar system. In this case the text needed to be processed *algorithmically*. In other words, it served as *input* for some program, whereas a new numerical text, again referring to (future) observations of the natural world, served as its *output*.

The invaluable advantage of the positional system was its ideal adaptability to algorithmic processing—in particular, to simple and universal rules of addition and multiplication that could be taught to schoolchildren and clerks. The more complicated programs—those intended for clerks—were described in natural language using an iteration of elementary algorithms along with conditional statements ("if the debit of client NN exceeds his credit by ZZ florins, stop deliveries").

The nascent programming languages for centuries existed only as informal subdialects of a natural language. They had a very limited (but crucially important) sphere of

Figure 1. Gregorio Reisch, *Margarita Philosophica*, Strasburg, 1504. Biblioteca de Catalunya. Arithmetic contemplates an abacist and an algorist.

applicability, and were addressed to human calculators, not electronic or mechanical ones. Even Alan Turing in the 20th century, when speaking of his universal formalization of computability, used the word "computer" to refer to a person who mechanically follows a finite list of instructions lying before him/her.

The ninety-page table of natural logarithms that John Napier published in his book *Mirifici Logarithmorum Canonis Descriptio* in 1614 was a paradoxical example of this type of activity that became a cultural and historical monument on a global scale (Coco Chanel's intuition did not deceive her). Napier, who computed the logarithms manually, digit by digit, combined in one person the role of creator of new mathematics and that of computer-clerk who followed his own instructions.

Deg. 0 + | —

mi	Sines	Logarith	Differen.	Logarith	Sines	
0	0	Infinite.	Infinite.	.0	1000000.0	60
1	291	8142567	8142568	.1	1000000.0	59
2	582	7449419	7449421	.2	999999.8	58
3	873	7043952	7043956	.4	999999.6	57
4	1164	6756275	6756274	.7	999999.3	56
5	1454	6533131	6533130	1.1	999998.9	55
6	1745	6350810	6350808	1.6	999998.6	54
7	2036	6196659	6196657	2.2	999998.0	53
8	2327	6063128	6063126	2.8	999997.4	52
9	2618	5945345	5945342	3.5	999996.7	51
10	2909	5839986	5839814	4.3	999995.9	50
11	3280	5744676	5744671	5.2	999995.0	49
12	3491	5657665	5657658	6.2	999994.0	48
13	3781	5577622	5577615	7.3	999992.8	47
14	4072	55[illegible]3514	5503506	8.4	999991.7	46
15	4363	5434522	5434513	9.6	999990.5	45
16	4654	5369984	5369973	10.9	999989.2	44
17	4945	5309360	[illegible]309[illegible]48	12.3	999987.8	43
18	5236	5252202	[illegible]252188	13.8	999986.3	42
19	5527	5198136	5198120	15.4	999984.7	41
20	5818	5146843	5146836	17.0	999983.1	40
21	6109	5098054	5098045	18.7	999981.3	39
22	6399	5051534	50[illegible]1514	20.5	999979.[illegible]	38
23	6690	5007083	5007060	22.4	999977.6	37
24	6981	4964524	4964499	24.4	999975.6	36
25	7272	4923703	4923676	26.5	999973.6	35
26	7563	4884483	4884454	28.7	999971.4	34
27	7854	4846743	4846712	30.9	999969.2	33
28	8145	4810376	4810343	33.2	999966.8	32
29	8436	4775286	4775250	3[illegible]	999964.4	31
30	8726	4741385	4741347	38.1	999961.9	30

Min.

Deg. 89

Deg. 0 + | —

mi	Sines	Logarith	Differen.	Logarith	Sines	
30	8726	4741385	4741347	38.1	999961.9	30
31	9017	4708596	4708555	40.7	999959.3	29
32	9308	4676848	4676805	43.4	999956.6	28
33	9599	4646077	4646031	46.1	999953.9	27
34	9890	4616225	4616176	48.9	999951.1	26
35	10181	4587239	4587187	51.8	999948.2	25
36	10472	4559069	4559014	54.8	999945.2	24
37	10763	4531671	4531613	57.9	999942.1	23
38	11054	4505004	4504943	61.1	999938.9	22
39	11344	4479030	4478965	64.4	999935.7	21
40	11635	4453713	4453645	67.7	999932.3	20
41	11926	4429022	4428950	71.1	999928.9	19
42	12217	4404925	4404850	74.6	999925.4	18
43	12508	4381396	4381318	78.2	999921.8	17
44	12799	4358408	4358326	81.9	999918.1	16
45	13090	4335936	4335850	85.7	999914.3	15
46	13380	4313958	4313868	89.6	999910.5	14
47	13671	4292453	4292360	93.5	999906.5	13
48	13962	4271401	4271304	97.5	999901.5	12
49	14253	4250783	4250682	101.6	999898.4	11
50	14544	4230583	4230477	105.8	999894.2	10
51	14835	4210781	4210671	110.1	999890.0	9
52	15126	4191364	4191250	114.5	999885.6	8
53	15416	4172317	4172198	118.9	999881.1	7
54	15707	4153627	4153504	123.4	999876.6	6
55	15998	4135279	4135151	128.0	999872.0	5
56	16289	4117263	4117130	132.7	999867.3	4
57	16580	4100664	4100527	137.5	999862.5	3
58	16871	4082175	4082032	142.4	999857.7	2
59	17162	40[illegible]5082	4064935	147.3	999852.7	1
60	17452	4048276	4048124	152.3	999847.7	0

Min.

Deg. 89

Figure 2. A table of logarithms.

All the more amazing was the philosophical insight of Leibniz, who in his famous exhortation *Calculemus!* postulated that not only numerical manipulations, but any rigorous, logical sequence of thoughts that derives conclusions from initial axioms can be reduced to computation. It was the highest achievement of the great logicians of the 20th century (Hilbert, Church, Gödel, Tarski, Turing, Markov, Kolmogorov, . . .) to draw a precise map of the boundaries of the Leibnizian ideal world, in which

- reasoning is equivalent to computation;
- truth can be formalized, but cannot always be verified formally;
- the "whole truth"—even about the smallest mathematical infinity, natural numbers—exceeds the potential of any finitely generated language to generate true theorems.

The central concept of this program, *formal languages*, inherited the basic features of both natural languages (written form fixed by an alphabet) and the positional number systems of arithmetic. In particular, any classical formal language is one-dimensional (linear) and consists of discrete symbols that explicitly express the basic notions of logic.

One crucial idea that laid bare unavoidable Gödelian "incompleteness" of any finitary deductive system consisted in incorporating the capacity for self-reflectivity of natural language (the "liar's paradox") into formal languages. This self-reflectivity is possible thanks to Leibniz's arithmetization of language.

* * *

Any concrete mathematical text includes words and formulas. With some stretch of imagination, the formulas can be regarded as expressions of a formal language (the language may change from one paper to another, but often it is simply a version of the language of set theory). The question of how words and formulas share the function of transmitting content is interesting in its own right. Most importantly, the words are what directs the paper toward people rather than card-readers; they provide unformalized short cuts, associations, motivation; they are charged with such subtle tasks as conveying the author's value system.

But in core fragments of mathematical text it is not always and everywhere formulas that are used to convey meaning.

At least since the time of Euclid—and in high school geometry textbooks in our time—diagrams have played the role of formulas. Most of us remember the drawing of a square which two lines divide into two smaller squares and two rectangles. This drawing illustrates/replaces/proves the formula $(a + b)^2 = a^2 + 2ab + b^2$.

A much more interesting and much less-well-known diagram (see Fig. 1) illustrates a classic theorem of Pappus of Alexandria (circa 300 A.D.). This drawing nicely exemplifies how the geometrical thinking of mathematicians has interacted with their formulaic and formal thinking over the course of many generations and centuries.

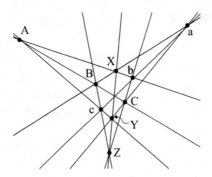

Figure 3. Pappus's construction.

Let us state the theorem first. We begin with a hexagon in the plane with vertices BXbCYc. (It does not have to be convex, as it is in the picture! This is the first pitfall of drawings—they often impose restrictions that we do not realize are there.) Any pair of opposite sides of the hexagon, say Bc and bC, determine a diagonal XY that passes between them. We extend these two sides and the diagonal; it might happen that the resulting three lines intersect in a point Z, in which case we say that the two sides have *the Pappus property*.

Pappus's Theorem. *If two pairs of opposite sides of a hexagon have the Pappus property, then so does the third pair.*

This is an amazing result. First of all, it is hard to imagine how one would arrive at this statement. Unlike the "$(a + b)^2 = a^2 + 2ab + b^2$" picture, it does not live in the realm of

Euclidean geometry: distances, lengths, and angles play no role whatsoever in the theorem or its proof, and neither does the group of rigid motions of the plane.

The required structural elements are primitive and strictly combinatorial: the plane consists of points, lines are certain subsets of points, two lines intersect in one and only one point, one and only one line passes through two given points.

It was only in the 19th century that it became clear that Pappus's theorem is a central result of *projective* plane geometry. It first seemed to be a statement about the geometry of the usual plane over the real numbers, completed with a line "at infinity," where parallels meet. But then it was found that the same is true for the projective plane over any abstract field. Moreover, the field with its composition laws and axioms can be recovered from the Pappus construction.

Finally, near the end of the 20th century, it turned out that the equivalence between Pappus's theorem and the theory of commutative fields can be explained and generalized in a broad setting by the *theory of models*. To put it simply, a *model of a formal language L* is a translation of L in the language of set theory, together with a standard interpretation of the latter. Thus, the meaning of the elaborate diagram of Pappus is revealed in a complicated metalanguage construction.

To summarize, model theory is a metalinguistic device that makes explicit a potential *intrinsic geometric content of language—like, linear, discrete, formal theories*.

But can one cross this bridge in the reverse direction?

Can one produce meaningful mathematics expressed "only" in pictures/geometric images...?

* * *

For a number of reasons pictures, such as Pappus's configurations, resist attempts to organize them into a language. Their inner syntax is idiosyncratic and not systematic; the syntactic links between them defy formalization. Any picture possesses a type of integrity that is lost in the process of analysis. The way they transmit and preserve information in various processes differs from the functioning of even "synonymous" language designs. They appeal to a different type of imagination—to our right-brain intuition.

Explaining the (presumed) intention of creators of cubist art, such as "Harlequin with violin," E. H. Gombrich writes:

> Critics considered it as insult to their intelligence to be expected to believe that a violin "looks like that." But there never was any question of an insult. If anything, the artist paid them a compliment. He assumed that they knew what a violin was like, and that they did not come to his pictures to receive this elementary information. He invited them to share with him in this sophisticated game of building up the idea of a tangible solid object out of the few flat fragments on his canvas.

Actually, these words describe very well certain key images and key formulas in the history of mathematics, including a charged relationship between them.

With the development of homological algebra and category theory in the second half of the 20th century, the language of commutative diagrams began to penetrate into mathematics; it took some time for mathematicians to get used to "diagram-chasing."

Figure 4. Pablo Picasso. "Harlequin with Violin" (1918).

Figure 7 shows an actual example of such a diagram (taken from a 2007 paper by D. Borisov and the author). The commutative square shown in the second part of Figure 7 is an elementary component of the diagram. Before the era of category theory, we could almost encompass the statement expressed by this square in linear language by writing the equality $h \circ f = k \circ g$. However, there is a caveat: f, g, h, k are morphisms in a category, and one needs to know what object each morphism comes from and what object it "hits."

Moreover, the large diagram in Figure 7 contains some slanted arrows, such as α. Such an arrow represents a morphism that is *not in the original category* $\mathcal{C}$ where the objects live whose names are given at the beginning and end of the arrows. Rather, they represent morphisms in the *category of morphisms M or* $\mathcal{C}$:

$$\alpha : Id \circ F' \longrightarrow F'' \circ G.$$

The exact content of a diagram can be conveyed in a linear text only through detailed commentary that alternates words with formulas. But does such a text render the diagram superfluous? No! (Once I was corresponding with a colleague by email about a very concrete mathematical subject. In email text one must, of course, find verbal equivalents for anything visual. Suddenly from my colleague I received a cry of the soul: "A diagram! My kingdom for a diagram!")

Below I will argue that the development of category theory and especially homotopic topology during the last few decades not only was a major step forward in a concrete area of mathematics, but also contributed to the realization and verbalization of an epistemological shift in our understanding of the foundations of mathematics that is occurring before our eyes.

Here I should clarify that for me "foundations" do not have any prescriptive or normative function. I understand "foundations" to mean the fruit of the labor of mathematicians as they decide which problems to tackle, how to write definitions and proofs, and, last but not

$$(a + b)^2 = a^2 + 2ab + b^2$$

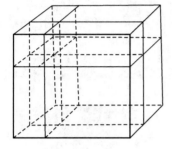

$$(a + b)^3 = a^3 + 3a^2b + 3ab^2 + b^3$$

(2-dim projection of) 3-dim cube

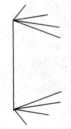

Newton's binomial formula

$$(a + b) = \sum_{i=0}^{n} C_n^i \, a^i \, b^{n-1}$$

2-dim projection of n-dim cube ??

Figure 5. Mathematical cubism.

least, how to transmit their knowledge, both inherited and newly discovered, to the next generation(s) of mathematicians.

The most important social function of research into foundations is probably to maintain a dialog between the "two cultures" (C. P. Snow). This dialog begins because mathematics continually causes a strong philosophical anxiety.

If one does not accept that a Platonic world of ideas exists independently of us (and philosophers sometimes do not even accept the existence of the world of objects and phenomena), then one has to regard mathematics simply as the fruit of the highly trained imagination of several dozens or thousands people in each generation. Leaving aside for now the question of the criteria for "truth" of a mathematical statement, one has to be astonished by the stability of mathematical knowledge and by its reproducibility across generations and civilizations.

Moreover, this knowledge is not just reproduced in the way that the texts of *The Odyssey*, the *Epic of Gilgamesh*, and *The Gospel* are. Over the last two centuries it has developed and become richer at a rate unheard of in earlier times, changing its form several times.

* * *

Archimedes's Tomb

$$\int_{-1}^{1} x^2\, dx = \frac{2}{3} \quad \Longleftrightarrow$$

A sphere has 2/3 the volume and surface area of its circumscribing cylinder. A sphere and cylinder were placed on the tomb of Archimedes at his request.

Figure 6. The sphere and the cylinder.

Let us return to the question of the mathematical content of the "foundations of mathematics" and its historical evolution over the last 150 years. For the great majority of mathematicians working after, say, World War II, the primary mathematical image they have is that of a set with some additional structure: topological spaces, groups, rings, measure spaces, ... At first, this set is just a Cantor abstraction: the nature of its elements is not important. What is important is that they are pairwise distinguishable and are conceptualized as parts of a whole. In the next stages, the elements of a new set might be the open subsets of an earlier set, the local functions on it, and so on.

Cantor himself, in a minimalist spirit, asked the most basic questions about these sets, demonstrated that there is an infinite scale of infinities, and left the problem of understanding the ontology and epistemology of this scale to the generations of logicians that followed him. The more pragmatic generation that survived World War I took this potentially metaphysical

Mathematics, Art, Civilization

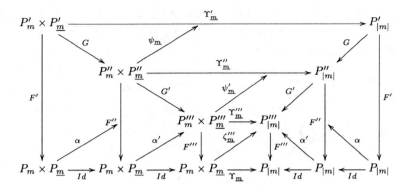

Commutative 2-diagram

$$A \xrightarrow{g} B$$
$$f \downarrow \quad \downarrow k$$
$$C \xrightarrow{h} D$$

Commutative square

Figure 7. Some commutative diagrams.

base and constructed from it an architecturally modern and functional building for the working mathematician, using the industrial products called "structures" in the sense of Bourbaki.

Working mathematicians put questions about the scale of infinities on the back burner; however, they kept the discrete sets as the basic construction material. The continuous became a superstructure placed over the discrete.

Even before Cantor, it was completely clear that there were problems with set constructions even in elementary arithmetic. If the natural numbers denote the number of notches on a stick or the cardinality of any finite set,

$$I, II, III, IIII, \ldots,$$

then the notion of zero as the cardinality of the empty set creates psychological problems, and negative numbers necessitate introducing either some artificial algebra or else an interpretation in a completely different realm, such as economics ("debt").

On the other hand, if we regard the continuous as the starting point of intuition and the discrete as a derivative structure, then the integers can be introduced in an extraordinarily natural way. Imagine a point moving in the plane. Suppose it starts at some initial position, wanders around for a while, and then returns to the starting point, without ever passing through the origin. We ask: How many times did it circle around the origin? It is not hard to give a precise definition of this integer, which can be zero, positive, or negative (since the point can circle around the origin clockwise or counterclockwise). Moreover, it is easy to see that a path that circles one way is canceled by a path that circles the other way—this is abbreviated as $1 - 1 = 0$. That is, the path obtained by putting these two together can be contracted to a point without passing through the origin.

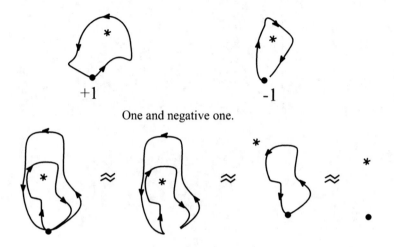

$1 + (-1) = 0$: the equality is replaced by homotopy

Figure 8. The picture of homotopy.

So what comes first, discrete or continuous? Certainly, it is an archetypal philosophical question: λόγος symbolizes discrete, and χάος continuous.

Using a metaphor from a nearby discipline, ethnography, I would compare this situation with Lévi-Strauss's theory of myth. Undoubtedly influenced by Bourbaki, Lévi-Strauss constructed his interpretation of myth as a mediation of opposites.

A quarter century ago I was thinking about this idea and conjectured an evolution in the reverse direction. According to this viewpoint, myth marks an epoch when awareness of opposites ("the discrete") was born of mental chaos. In this way musical notation was born from the music itself.

In the mind's eye, the geometer who works in homotopic topology sees infinite dimensional spaces that can and must be deformed until they contract to a point. In the last analysis, the discreteness that the topologist computes and conveys in a discrete language reduces to the "connected components" of these spaces and the induced spaces of mappings.

In popular expositions of mathematics—and now in videos as well—"knots" in $\mathbb{R}^3$ and "turning a sphere inside out" are used to externalize these private mental images. The possibilities of such externalizations for pedagogical purposes are as limited as the possibility of imagining oneself to be Sviatoslav Richter performing Schubert after seeing him interviewed by Bruno Monsaingeon.

For this reason I can only briefly describe my impressions of the epistemological shift that I see making itself felt in the foundations of mathematics. Its essence is a reversal of the relation between the discrete and continuous, between language and imagination, between algebra and topology. Continuity, topology, the geometrical imagination—these slowly take over the place once occupied by elementary mathematics.

Figure 9. Manuscript of Sonata for Violin Solo, by J. S. Bach.

The language becomes secondary, subordinate, its "internal writing" returns to the archaic hieroglyphic form, and the combinatorics of geometric images becomes its content. This combinatorics is nonlinear and multidimensional, and even at birth the new language mixes syntax, semantics, and pragmatics in ways that we have not yet begun to understand philosophically.

The commutative diagrams of category theory were a foretaste of this evolution. With the advent of polycategories, enriched categories, A_∞-algebras, and similar structures, we are beginning to speak a language that is much less amenable to externalization than what we are used to.

For me a very convincing argument that this perception is more than just my personal fantasy was the realization that there are parallel processes on the border between mathematics and theoretical physics. I have in mind Feynman integrals, renormalization methods, and such applications as Witten's integral for calculating knot invariants.

* * *

I want to conclude by returning to the theme I started with: the problem of the persuasiveness of mathematics and, more generally, modern science.

Personal experience, eyewitness accounts, references to authorities or authoritative texts—these are often taken as a complete list of the means of persuasion. Of course, physicists, chemists, and biologists would add experimental evidence to this list.

But I would like to consider as well what I call a "civilization" argument—the argument that Coco Chanel grasped intuitively. Civilization provides us with means to verify truth that do not reduce to appeals to authority, to witness testimony, or to reading through lengthy mathematical proofs.

When preparing this essay I made extensive searches on the Web—the possibility of this kind of electronic communication is taken for granted now by almost everyone. But it was made possible by two thousand years of mathematics the full persuasiveness of which cannot be verified by us or anyone we regard as authorities.

Among other reasons, mathematics is true because the discovery of Maxwell's equations led to the technique of transmitting information by electromagnetic waves, and Boolean algebra is hard at work in your and my laptops.

The culture of mathematical reasoning in its "civilization" aspect is the most important way to objectify abstract mathematical knowledge and transmit it from generation to generation.

On a personal level I would compare mathematical culture—the culture of proofs—with a musician's training: working out the details of small movements until they become automatic and can be synthesized, say, in Bach's "Sonata for Solo Violin."

The codification of formal language, with its components of logic and set theory, was the ideal means for "working out the details of small movements." But if it is accompanied by ideological propaganda, such as intuitionism or constructivism, it becomes philosophically weakened and loses its "civilization" value.

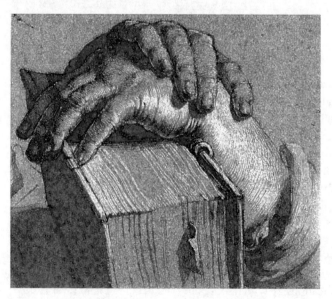

Figure 10. Dürer, hands on book.

18

Questions about Mathematics

Harold R. Parks

If a question can be put at all, then it can also be answered.

—Wittgenstein

Why Do Humans Even Know That Mathematics Exists?

Archaeological evidence shows that anatomically modern humans appeared about 200,000 years ago. Fully modern human behavior (figurative art, music, trade, and so on) dates back at least 30,000 years. There are two theories regarding the emergence of modern behavior. One is that knowledge, skills, and culture gradually accumulated to a sufficient degree that the behavior of *Homo sapiens* crossed the metaphorical boundary into modern behavior. The second theory for the emergence of modern behavior is called the Great Leap Forward or, less colorfully, the Upper Paleolithic Revolution. This theory posits a genetic change that led to the development of spoken language and, consequently, to modern behavior. The assumption that a genetic change occurred as required by the Great Leap Forward theory may not be as *ad hoc* as it seems. In fact, there is independent, scientific evidence of a genetic change related to brain size regulation that occurred at approximately the time of the Great Leap Forward (see [EV]).

Written language was not developed until about 5000 years ago—3000 BCE—so we can only speculate as to what humans were thinking about before that time. Presumably personal and communal survival were the predominate interests, but the existence of art, music, and burial rites argues for some abstract thought. Whether their abstract thought was as deep as ours—and specifically whether any of those thoughts had mathematical content—we cannot know. Complicating matters is the fact that further genetic changes related to brain brain size regulation occurred as recently as 3800 BCE (see [ME]), that is, less than a millennium before the development of written language.

It is believed that written language was developed as the result of a gradual process that began with the need to keep track of stored agricultural goods. The first way of recording what was being stored was to make clay tokens to represent particular quantities of the various items. Ownership was indicated by enclosing the tokens in a spherical clay envelope marked on its surface to show who owned the contents. Since clay is not transparent, one has the difficulty that you don't know what is in the envelope unless you break it open—but then you need to make another envelope. To get around that difficulty, the tokens were impressed upon the clay of the envelope before being sealed inside. Eventually people realized that the impressions alone suffice to do the record keeping and that the impressions may as well be made on a flat clay tablet as on the surface of a sphere—in essence a written record has been made.

We see that the primitive mathematics of recording numbers arose contemporaneously with the development of civilization and led to the development of written language. As civilization grew more complex, so did the complexity of the mathematics used, moving beyond mere counting to include place-value notation and arithmetic. Indeed there is solid evidence that the notion of compound interest was understood as early as 2400 BCE. For more on these matters, see [SC].

Similarly, it is believed that geometry developed from practical needs. The traditional story is that land management in Egypt was the driving force behind the development of geometry. The source of that traditional story is the Greek historian Herodotus (484–425 BCE). Herodotus relates the following about the Pharaoh Sesostris—a legendary figure, apparently based on several Pharaohs including Senusret III who ruled in the mid 1800s BCE (see [HE]):

> Sesostris also, they declared, made a division of the soil of Egypt among the inhabitants, assigning square plots of ground of equal size to all, and obtaining his chief revenue from the rent which the holders were required to pay him every year. If the river carried away any portion of a man's lot, he appeared before the king, and related what had happened; upon which the king sent persons to examine, and determine by measurement the exact extent of the loss; and thenceforth only such a rent was demanded of him as was proportionate to the reduced size of his land. From this practice, I think, geometry first came to be known in Egypt, whence it passed into Greece.

We have seen that, historically, mathematics was developed for its practical uses. The focus was on doing calculations and teaching people how to do calculations. Up until the wide availability of low cost pocket calculators in the 1980s, arithmetic calculation remained the focus of elementary school mathematics. Of course, it is always enjoyable when you can be the best at something, so I thought grade school arithmetic was tremendous fun.[1] I expect that it would be an unusual mathematician of my generation or any earlier generation who as a youngster did not excel at and enjoy arithmetic calculations in grade school. On the other hand, there are a lot of people who were very good at arithmetic, but who did not become mathematicians. There is a gulf between doing practical calculations and thinking deeply about mathematics. We would hypothesize that, sometime after the invention of written language (circa 3000 BCE), some group was inspired to look into

---

[1] I make no claim to ever having been a great arithmetician: I attended small rural elementary schools, so the level of competition was low.

the deeper patterns of mathematics. It had been thought that the clay tablet known as Plimpton 322, dating to 1800 BCE and containing a list of Pythagorean triples, gave clear evidence that that transition had occurred by then. Recent scholarship has undermined that theory. In [RO], Robson argues convincingly from the language and historical context, as well as the mathematics, that Plimpton 322 was more likely a teaching aid rather than the product of deep mathematical insight. We are left to wonder when the transition from doing calculations to thinking deeply about mathematics occurred. Certainly by the time of Thales of Miletus (circa 640 BCE–546 BCE) the transition had been made. We know this with certainty because Thales used the deductive method in geometry.

Perhaps the remarkable thing is that the transition did happen. You can't eat mathematics—it's no direct help with subsistence. On the contrary, sometimes losing oneself in mathematical thought can be maladaptive. A friend described how he was a menace as a construction worker, because he was so often lost in thought about mathematics; I believe him because I once witnessed him sweeping a mop back and forth, endlessly and pointlessly, through the same puddle of spilled wine, while he focused his attention on the mathematical discussion in which he was participating. Nonetheless, over the past two or three thousand years, quite a number of otherwise sane people have become entranced with mathematics to the point of devoting much of their lives to deep study of abstract mathematical concepts. I am led to conclude that there is something about the human brain that, in some people, allows and sometimes compels an intense interest in the abstract patterns of mathematics. If the human brain was evolving 5800 years ago, it may still be, but whatever it is that makes some of us mathematicians, it has existed for at least 2700 years.

What Leads Some to Choose Mathematics, While Most Do Not Make That Choice?

To be a mathematician, you need education, interest, and ability. If we look at the entire world population over the last two and one half millennia, most had no opportunity to even learn to read and write. In the modern developed countries, we take universal, free (i.e., tax supported) education for granted. In fact, this access to a free education came rather late. In England, it was not established until 1870 when Parliament passed the Elementary Education Act. In the United States, elementary education was a matter for the states, so progress was uneven. As a proxy for universal free education we can use compulsory school attendance laws: The first such law was established in Massachusetts in 1852. Of the fifty states, seventeen did not have compulsory school attendance laws until after 1900. Alaska was the last state to enact a compulsory school attendance law (in 1929).

Much of the available education in the United States in the 1800s must have been of questionable quality, taught by the barely qualified. A glimpse into the nature of education in the United States in Ohio in the mid 1800s is available to us from an unlikely source: the memoirs of President U. S. Grant. Grant tells us (see page 19 of [GR]),

> The schools, at the time of which I write, were very indifferent. There were no free schools, and none in which the scholars were classified [i.e., separated by grade].

> They were all supported by subscription, and a single teacher—who was often a man or a woman incapable of teaching much, even if they imparted all they knew—would have thirty or forty scholars, male and female, from the infant learning the ABC's up to the young lady of eighteen and the boy of twenty, studying the highest branches taught—the three R's, "Reading, 'Riting, 'Rithmetic." I never saw an algebra, or other mathematical work higher than arithmetic, in Georgetown [Ohio], until after I was appointed to West Point. I then bought a work on algebra in Cincinnati; but having no teacher, it was Greek to me.

The mode of instruction in U.S. schools of the day was barbaric, for Grant tells us (see page 23 of [GR]),

> I can see John D. White—the school-teacher—now, with his long beech switch always in his hand. It was not always the same one, either. Switches were brought in bundles, from a beechwood near the school-house, by the boys for whose benefit they were intended. Often a whole bundle would be used up in a single day.

Education at the college and university level in the United States was also questionable by our current standards. Dartmouth College, from which I graduated in 1971, is a well regarded liberal arts college with a strong undergraduate mathematics program, but in the 1800s students at Dartmouth were required to study math for only their first two years. Apparently the mathematics instruction did not generate much enthusiasm among the students because, beginning sometime in the 1830s, Dartmouth sophomores would celebrate the end of their math studies by burying their textbooks in a pseudo funeral.

Harkening back to President Grant again: he never calls himself a great scholar, but the fact that his performance in mathematics was better than in other subjects was sufficient justification for him to aim for a future as a professor of mathematics (page 27ff of [GR]).

> A military life had no charms for me, and I had not the faintest idea of staying in the army even if I should be graduated, which I did not expect. . . . Mathematics was very easy to me, so that when January [1840] came, I passed the examination, taking a good standing in the branch. . . . My idea then was to get through the course, secure a detail for a few years as assistant professor of mathematics at the Academy, and afterwards obtain a permanent position as professor in some respectable college; but circumstances[2] always did shape my course different from my plans.

In the present day United States, we have universal, free education available. The teachers at all levels are well educated. State boards insure that K–12 teachers have teacher training—though when you teach in a university you may doubt the efficacy of that training and teaching. In principle, everybody in the U.S. has the opportunity to prepare for higher education.

Not everyone takes advantage of the education offered free in K–12 and not everyone who could go to college does so. Attitude is a huge factor. A lot of people think they can't do it. A lot of people think they don't need it. My paternal grandfather was a chemistry

---

[2] The "circumstance" referred to is the start of the war with Mexico.

professor at Penn State, my paternal grandmother had a liberal arts degree, and my father and his sisters were all physicians. The role models were there. I went to a prep school, my grades and college board scores were high, and my family could pay for any college in the land. There was no question whether I would go to college, the issue was where. Nonetheless I was worried about how hard college would be. If *I* could be worried about college, it's no wonder a lot of people doubt their ability to succeed in college. There must be many people who could have been mathematicians if they had had more confidence.

Education is the first piece of the picture, interest is the second. Not everybody is interested in the same things. I had a childhood acquaintance who, when he was about 12 years old, decided that he wanted to be a dentist. Over the years our paths would cross occasionally, and we would chat a little. I ran into him during our college years and sure enough he was in the pre-dental program at Penn State. Then, when I was in graduate school, he was in dental school. He succeeded; he did become a dentist. That's great. I'm happy for him.

I don't know how the topic came up, but I once related the preceding little tale about my friend-who-became-a-dentist to my wife. At the end I added as an editorial comment, "Well OK, dentistry is a respectable profession, but man, talk about boring!" That was the punch line of a joke I didn't get, because my wife cracked up laughing, while I just stood there with a goofy look on my face. When I asked her what was so funny, she said, "What do you do?" Oh yeah, most people, if they thought about it at all, would be thanking their lucky stars each Monday morning that they don't have to trudge to the math department to face another long week of mathematics.

Interest in mathematics is not universal. Early experiences with poor teaching can stunt or kill any incipient interest. Another friend who is about my own age told me that, when she was in fifth grade, the teacher locked her in the classroom's closet to force her to study the multiplication table, which she had failed to memorize. This did not improve my friend's attitude toward math. Perhaps teachers of mathematics also should have a precept like the physicians' "First, do no harm."

On the other hand, good teaching can inspire interest. I try to make inspiring interest in and enthusiasm for mathematics a high priority. Of course, as Burns wrote, "The best laid schemes o' mice an' men gang aft agley." I was once asked to give a short survey talk about my research to a group of prospective women students. The goal was to generally encourage interest in studying mathematics among women. There are a lot of visual things that I can present in such a talk concerning my work on area-minimizing surfaces, and specifically, computing approximations to such surfaces. So, as long as I avoid the technical details, it can be a reasonable thing to present to a general audience. Everything went great. The students seemed to enjoy it and be interested. Naturally questions were welcomed. Finally one of the women asked, "How long have you been working on this?" In fact, investigating one idea for computing area-minimizing surfaces had been my thesis topic. When you came right down to it, I'd been working on that general topic since a few days after I passed my qualifying exams though, of course, not exclusively. So after thinking it through, I answered honestly, "Twenty years." You could almost hear the enthusiasm evaporate as the women considered the prospect of working on this problem, or any other problem, for twenty years.

How Does One Become a Mathematician?

The standard path to becoming a mathematician is through the process of getting a Bachelors degree, Masters degree, and finally a Doctorate. To make it the whole way inevitably takes above average intelligence and above average perseverance. I do have a colleague who claims to not have high intelligence, but his claim only makes me doubt the accuracy of the testing he received.

Persistence is especially important because there are a lot of hoops to get through. One of those hoops may be the language exam(s). For a long time, as part of their Ph.D. programs, universities in the United States routinely required reading knowledge of one or two languages other than English. In the nineteenth century, most of the world's scientific research was taking place in Europe, so the requirement made sense.

For instance, my paternal grandfather had a Ph.D. in chemistry, and it seems plausible that in the 1920s he would have found a working knowledge of written German quite useful. On the other hand, it's not clear that my grandfather ever did attain that useful working knowledge of German. He told me, that during his Ph.D. program, he studied German under the tutelage of one of the chemistry professors, let's call him A. When my grandfather asked A if it was time to face the German language examiner—another chemistry professor—A said words to the effect, "You're not really ready, but give it a try." The language examiner's first question was, "Who have you been studying German with?" When grandfather named A, the examiner said, "That's good enough for me!" and thus my grandfather passed his exam.

Fortunately for native speakers of English, it is English that has become the dominant language of science, so the need to know French, German, or Russian has eased. Consequently, language requirements are gradually being eliminated from mathematics Ph.D. programs in the United States. Most current and future students will be spared the aggravation I went through trying to learn to read French without ever having even one formal lesson about the language[3] and with only the aid of Lusin's *Leçons sur les ensembles analytiques et leurs applications* and the cheapest French–English dictionary I could find.

Another major hoop is the exam(s) called variously the qualifying exam, the preliminary exam, or at Princeton, the general exam. Studying for these exams is a classic hurdle for students. I recall a student of mine who aspired to a Ph.D., but when it came to studying all summer for the qualifying exams in September, he said he just couldn't get motivated. Keeping the motivation to study hard over a prolonged period for one exam is difficult. If someone could put motivation into pill form (preferably non-addictive, non-toxic, and long-lasting), they'd get rich and, more importantly, many more students would pass the quals.

The hardest step in the process is writing a thesis. Some advisors provide their students with good problems. After I passed my general exam, I went to Fred Almgren to ask him to be my advisor, and he was ready with a collection of possible ideas to work on. Fred Almgren was also very positive and encouraging, so I was blessed in being able to work with him. Not everybody is so lucky in their choice of advisor. And, even when the start goes great, the middle and end are going to be a challenge. Most people will need to put in

---

[3] A girlfriend did explain how to say "*Comment allez-vous*," but that was too absurd to comprehend.

a lot of hard work, much of it exploring ideas that don't succeed. Of course, you want your ideas to succeed, but it is important to be your own worst critic and try to find any mistakes that you may have made. You also need to talk to others about your work. Explaining something to someone else almost always helps you to understand it better.

For myself, I like a systematic and organized approach so, as I worked on my thesis problem, I recorded everything I did in spiral bound notebooks. Spiral bound notebooks are good because pages don't fall out by accident. To get one out you must rip it out—something I save for really misguided work. I am still using spiral notebooks, but after forty years there are bunches of them stored in disparate places.

The key thing is to keep at it. Don't give up. Refuse to let other responsibilities keep you from working on it.

Ultimately, you will need to have a well-written and typeset thesis. Turning thoughts and notes into TEX (a widely used computer typesetting system for mathematics) requires some effort, but that is not the heart of the matter. Having valid arguments is the heart of the matter. One tool for determining whether an argument is valid is to write it down in *full detail*. Of course, writing arguments out in full detail is tedious work. So do it now! It won't be easier later. And really strive for full detail: I have found that one of the best ways to locate areas in my own work where there may be trouble is to do a search on the word[4] "clearly."

What Does One Do After Getting a Ph.D. in Mathematics?

The straightforward career choice is teaching at a college or university. There are other possibilities, but since I chose the university teaching path, I know little about them and can't offer advice about them. If you have already been teaching in graduate school and know you don't like it, then you would be doing yourself a favor not to get a teaching job. I had negligible teaching experience in graduate school, so it was only by doing it in my first postdoctoral position that I found out how I felt about it: I'm nervous in front of each new group of students; after the first week of a course, it's kind of fun to lecture; grading is unpleasant.

The United States has an enormous number of colleges and universities spread all over the fifty states. If you want a job, then be willing to go anywhere. Of course, you should take the most attractive offer you get, but that may or may not be somewhere you think you really want to go. After you're there for a year, you'll know whether the place is as bad or as nice as you thought it would be. If you find you really are in some hellhole, move on.

The year before I came on the job market in the 1970s after my first postdoctoral position, one of the graduate students there had limited his job search to the northeastern United States. The consequence was that he got no offers, was forced to widen his search, and finally ended up with a one-year visiting position at the University of Alaska at Fairbanks. I know Alaskans love Alaska, but for a man from the northeast who wanted to stay in the northeast, that was an undesirable outcome. I took that graduate student's experience as

---

[4] The joke here is that, when a mathematician says "clearly," he/she frequently does not know what he/she is talking about. Obfuscation can be the order of the day.

a cautionary tale, hence the first person to whom I gave the "be willing to go anywhere" advice was myself.

The job market in academic mathematics seems to have been terrible for decades. But that is statistical. You personally need only one job. In fact, you are generally limited to one job. While it is common and acceptable to take a leave of absence (if you can get it) from one academic job while trying out a new one, people have gotten into considerable trouble for receiving two salaries simultaneously from two institutions for simultaneously filling two full-time jobs.[5]

As you move up the ranks, the projects you work on often will be long term things. It can be hard to keep at it. But it is essential that you do so. Since motivation is still not available in pill form, it follows that, to be a successful academic mathematician, you need to generate motivation from within yourself. The promotion process for earning tenure can substitute metaphorically for someone holding a gun to your head, but once you have tenure the penalty for not publishing is lessened. To counteract the paucity of external rewards, you should take advantage of opportunities to give yourself some positive reinforcement: Every success should be appreciated and celebrated. Never minimize the importance of a success—a paper accepted, a raise, an award—all of them should be celebrated.

What Do We Do About Teaching?

First, we should all do the best we can at teaching. Of course, if you are at a research university, then your focus for advancement must be on research—only the most superbly gifted teacher can rely on teaching for advancement in a research university.

Since each of us already has a personality, we need to fit our teaching styles to our personalities. I tend to find lots of things amusing—if perhaps in a cynical way—so I find it natural to try to inject a bit of humor into my classes. That does waste some valuable class time, but my hope is that some students might come to class who otherwise would not if the presentation was drier. Wishful thinking you may say. To the contrary, I had a student who brought his girlfriend (who was visiting from another city) to one of my linear algebra classes. She was not a mathematics major, so she didn't follow the technical part of the class, but she found the whole show amusing to the point of laughing aloud. She enjoyed it so much that she asked her boyfriend to bring her along to my class the next time she came to visit him.

Of course, most students are not interested in making mathematics a career. Many only take the math classes that are absolutely required—see the math funeral at Dartmouth above. So overall the attitude may be bad, and that bad attitude can sour your own attitude. Also, as you become older, the gulf between you and the students becomes wider. The following quotation can be found in [PA].

> The children now love luxury; they have bad manners, contempt for authority; they show disrespect for elders and love chatter in place of exercise. Children are now tyrants, not the servants of their households. They no longer rise when elders enter the room. They contradict their parents, chatter before company, gobble up dainties at the table, cross their legs, and tyrannize their teachers. (Attributed to Socrates by Plato.)

---

[5] You can Google it for recent cases, but it's not a new thing.

The source listed is as it is given in [PA]; unfortunately no specific primary source is cited in [PA]. Despite its possibly apocryphal nature, this quote may ring true. I don't buy it. I think people remain the same as they have been for the last 2700 years. Without a doubt culture changes, but the underlying human nature does not.

The culture among students is a changing thing, and the culture among faculty evolves along its own path, with each surely affecting the other. Since we can't stop it, we need to live with it. If we're fortunate, we can enjoy it and laugh at it.

Apparently drug use was prevalent among college students thirty years ago. One of my first calculus students at Oregon State got in trouble for drugs back then. Naturally, he wanted to avoid prison, so he asked me to serve as a character reference for him and write a letter on his behalf. The story about the raid on the house where he lived had been in the newspaper, so I knew a little bit about what had happened. The newspaper article mentioned that the police had found hallucinogenic mushrooms in the refrigerator in the house. The student was upset about the police finding those mushrooms. *That* should never have happened because, as he said, "If I had had any idea that the mushrooms were there, I would have eaten them."

In that era of drug use, it seemed to some young faculty members, such as myself, that weak students were sliding through advanced calculus and as a result the standards for higher level courses were falling, too. As young idealists, we took it upon ourselves to clean up the mess. We raised the intellectual level of the work required in advanced calculus; we read the homework papers closely, and those papers were graded and returned promptly. No more sliding through. In fact, a significant fraction of the students in advanced calculus would not even bother to turn in a paper when an assignment was due. On the one hand, this was very disappointing. On the other hand, I could be righteously indignant about the laziness of the students who didn't turn in papers and it didn't require any reading on my part.

A few years later, I was back to teaching advanced calculus. The student culture had changed. Every student turned in a paper, and I had to read all of them. I was glad they were taking the course seriously, but I had a lot more work.

I may be the odd one out on the topic, but I can't agree with the sentiments attributed to Socrates. Over the course of my career at Oregon State, the students seem to have become more serious about their work. I hope this observation has some broader validity, and if it does, that the improvement continues. Alas, the pendulum is probably due to swing the other way. But, whatever happens, my motto is "Teach the students you have, not the students you wish you had."

References

[EV] Patrick D. Evans, Sandra L. Gilbert, Nitzan Mekel-Bobrov, et al., *Microcephalin*, a gene regulating brain size, continues to evolve adaptively in humans, *Science* 309(2005), pages 1717–1720.

[GR] U. S. Grant, *Personal Memoirs of U.S. Grant*, introduction by B. D. Simpson, University of Nebraska Press, 1996.

[HE] Herodotus, Herodotus: The Persian Wars, translated by George Rawlinson, in *The Greek Historians*, edited by F. R. B. Godolphin, Random House, 1942, page 132.

[ME] Nitzan Mekel-Bobrov, Sandra L. Gilbert, Patrick D. Evans, *et al.*, Ongoing adaptive evolution of *ASPM*, a brain size determinant in *Homo sapiens*, *Science* 309(2005), pages 1720–1722.

[PA] William L. Patty and Louise Snyder Johnson, *Personality and Adjustment*, McGraw-Hill, 1953, page 277.

[RO] Eleanor Robson, Words and pictures: new light on Plimpton 322, *American Mathematical Monthly* 109(2002), no. 2, 105–120.

[SC] Denise Schmandt-Besserat, Oneness, twoness, threeness: How ancient accountants invented numbers, in *From Five Fingers to Infinity*, edited by Frank J. Swetz, Open Court, 1994.

19

A Woman Mathematician's Journey

Mei-Chi Shaw[†]

*Dedicated to the memory of my parents,
Chiang Shaw and Lan-Fang Liu*

1 The Village of Righteousness

I was born in 1955 in Taipei, Taiwan. My parents were among the two million mainland Chinese who migrated to Taiwan in 1949 after the Nationalist government lost the civil war to the communists. My father was in the retreating Nationalist Air Force. At birth, I had three brothers and one sister, and two years later my younger brother was born. They named me after their favorite theater in Shanghai (Mei-Chi Grand Theater still exists) which they frequented. My parents always fondly remembered Shanghai, where they lived with my two eldest brothers from 1945–1947, after the war with Japan. They always said jokingly that, since girls do not count in a Chinese family, they did not give much thought to the naming of girls. My sister's name is Mei-Som—Mei means beautiful and Som means a special kind of jade. Mei-Chi means another kind of beautiful jade. All the boys were named in the family tradition with the word Yu in their name, meaning Universe (Fig. 1).

Taiwan in the 1950s was a very unusual place. The island had been under Japanese rule for fifty years, from 1895–1945, and was returned to China only after the war. Four years later, two million mainlanders from all over China fled with the Nationalist government to Taiwan to join the six million Taiwanese. We grew up on a military housing project near the Air Force headquarters, very much segregated from the outside world. At that time there were many government military housing compounds like this to accommodate the hundreds of thousands of Nationalist military personnel and their families who had retreated from the mainland to Taiwan. The housing compound where we lived was called the "Village of Righteousness." It was called a "village," but it was near the center of Taipei. Other villages had names like the "Village of Loyalty," "Village of Recovering," and "Village of

---

[†] Department of Mathematics, University of Notre Dame, Indiana, U.S.A. E-mail: shaw.1@nd.edu

Figure 1. 1961 Taipei Botanical Garden with my parents, elder Brothers Der-Yu, Zhen-Yu, Cuen-Yu, younger brother Chang-Yu and elder sister Mei-Som.

Reconstruction" to reenforce the government doctrines. When I was young, the communist threats were still very real and we were bombarded by government propaganda to be prepared for the communist invasion and to prepare for recovering and reconstructing the mainland next year (always next year).

All the people living in our village were Air Force families. These people came from all over China and spoke various Chinese dialects. Every family had a story to tell about how they came to Taiwan during that fateful year of 1949. My father told of how he arranged to get my mother and my two eldest brothers out of Hunan on the last flight from the Hen-Yang airport near my mother's hometown Leiyang, when the communists were already nearby. My mother's was how sad and reluctant she was to leave her ancestral home that day, accompanied by my uncles and other relatives (and servants), to meet with my father for the flight to Taiwan. Little did she know that it would be another thirty-eight years before she would see her hometown again.

The houses in our village were barracks built in haste under the very spartan conditions of a military in retreat. With six children, it was very hard for my parents to make ends meet on my father's meager salary. It was also difficult for them to adjust to the new life after moving to Taiwan with all the uncertainties. But, for the children growing up in the military compound, the place was like an endless summer camp. I have some of the fondest memories of my childhood of playing, to my heart's content, with my siblings and neighbors' kids. Whenever our parents would leave the six of us at home alone, we turned the place upside down playing hide-and-seek. I never knew what loneliness was then.

One of my fondest memories was flying kites in the autumn with all the village kids in the rice paddies. Even though the village was near the center of Taipei right next to the Air Force headquarters, on the other side there were still rice fields. After the fall harvest, the farmers did not mind people walking on the narrow ridges along the paddies. We would use anything available to us to make our own kites, starting from splitting bamboo for the kite frames, to pasting old newspapers on the frames. There was no money to waste on kites. We would compete to see whose kite was the prettiest and whose kite would fly the farthest. If the line broke and the kite flew away, all the village kids would run after the kite to retrieve the valuable threads. We would get a good scolding if we did not bring the threads back home to our mothers. At that time, material things were so scarce, everything was precious, even a spool of thread.

The other thing I remember was Chinese New Year. The people in the village would celebrate it with great fanfare for the whole fifteen-day period. After the big New Year's Eve feast, there would be firecrackers all night. We kids would always stay up all night playing poker with the precious money we just received from our parents. The last day of the New Year celebration (the fifteenth of the first month of the lunar calendar) was the Lantern Festival. All the kids in the village celebrated by making our own lanterns together from scratch. My eldest brother was very handy and always made the most elaborate lanterns by hand. The traditional food for the Lantern Festival was sweet dumplings filled with sesame paste. After a big meal finished with the dumplings, all parents would turn out the lights in each house. We kids would carry our home-made lanterns, lit with candles, to roam the village in complete darkness. There must have been hundreds of kids together. We would form a long line like a dragon with the oldest ones in front and kids as young as two at the tail end. All the kids would chant together, "Here come the lanterns! Here come the lanterns! Dong, dong, dong," repeatedly until midnight. We did not want to go home until our parents started calling each kid to come home or face punishment. I always remembered how sad I was when I went back home that day thinking, another 350 days until the next New Year! These memories seem so distant now. Within twenty years, Taiwan would be transformed from an agrarian society into one of the four tigers in Asia, the others being Hong Kong, Singapore, and South Korea. The rice paddies near the village became the choicest real estate in Taipei and our village was also transformed into high rises. Military villages became a thing of the past.

By the time I was five, most other kids of the same age would attend the kindergarten on the compound. But I did not want to go since I would rather be playing outdoors than sitting in a classroom. My parents did not force me to attend, nor did they teach me how to read or write at home before I entered elementary school one year later. Unlike most other parents, they did not believe in early education to make kids get ahead, but believed that kids should play at that age. Though they were quite strict in disciplining children, they let us do things that we liked. One day my mother took me to see a fortune teller at a neighbor's house. The neighbor, who was very much into such things, highly recommended this fortune teller. Seeing a mother with a precocious five-year old daughter, the fortune teller started confidently by saying that my mother has no sons. After the neighbor corrected him, he said that originally my mother was destined to have no sons, it was only because she had done many good deeds in her previous life that she was blessed with sons in this life. The fortune teller thus successfully rescued himself from a misstep. He continued to compliment my mother: with the good physical features that she had, good fortunes were waiting for her around the corner, even though she had gone through a lot of hardship earlier. Such words could have been applied to and would have been welcomed by almost every parent in our village. At the end, my mother casually asked the fortune teller to also tell the fortune of her daughter. The fortune teller said to my mother, probably from a well-rehearsed line:

Your daughter will grow up to be smart and strong, better than men.

My mother was elated! She would repeat these words to me (almost) every day until I left home for graduate school. She always said to me how accurate that fortune teller was. I always reminded her that the same fortune teller also said that she had no sons.

2 My Education in Taiwan

2.1 Air Force Affiliated Primary School

I attended the elementary school affiliated with the Air Force in Taipei, located right next to the Air Force headquarters. There was an entrance gate directly linked to the headquarters with a plaque that read, "Air Force Children's Primary School." Under it there was a sign that proclaimed, "Founded in Jian-Qiao, Hang-Zhou." The China Air Force was founded in Jian-Qiao, Hang-Zhou before the war. Due to its remote location, a primary school was established in 1934 for the education of the children of Air Force personnel. Subsequently, it had moved several times, first to Chengdu, Sichuan during the Sino-Japanese war and then to Nanjing before it moved to Taipei in 1949.

Like many things in Taiwan, this children's primary school, was a refugee version of the original one in Jian-Qiao, Hang-Zhou, in mainland China, but the principal was the same. There were about a dozen such primary schools affiliated with the Air Force in Taiwan. All the children of Air Force personnel could attend for free, regardless of their rank. Among my classmates, there were children of generals and children of cooks. It was a co-ed school and with no uniform, since not every student could afford uniforms. All the students, regardless of their provincial ancestry and despite the fact that the official language was Mandarin, conscientiously adopted the Sichuan dialect as the common language among ourselves, even though dialects were banned by the teachers at school. At home, we children also spoke the Sichuan dialect to each other, even though only my eldest brother was born in Chengdu. My parents spoke their heavily accented Mandarin to us. It was quite a phenomenon that students of a primary school in the center of Taipei, born in Taiwan by parents from all over China, would continue to speak Sichuan dialect. In later years, I have had great difficulty explaining to people why the Sichuan dialect was my first language, both to people in Taiwan or from China. The good side of this development is that the diverse and isolated (refugee) students developed a special pride and bond from the strong identity the language, almost a secret code, offered. The downside was that many students did not adjust to the outside world. After I graduated from the elementary school in 1967, compulsory education in Taiwan was extended from six years to nine years. Taiwan's economy also took off in the 1960s. The Air Force Primary Schools all became public primary schools. Nevertheless, I still frequently encounter people from Taiwan with a Sichuan accent—and, if they are of my age, I can guess very often correctly where they went to primary school.

On the first day of primary school, the teacher asked what my name was and how to write it. I told her that I did not know how to write. Most other kids had attended the kindergarten and knew how to write their names already. My last name Shaw (Xiao) consists of nineteen strokes and is one of the most difficult Chinese characters to write. I remembered hearing my teacher commenting to another teacher: this is the daughter of Chiang Shaw, but she cannot even write her name! My father was the editor-in-chief of *China Air Force*, a monthly magazine, and had some reputation in writing. When I went home that day, my father wrote my name on a piece of paper and told me to memorize it. These were the first characters I learned when I was already six years old. I learned several characters in school each day for the next few years. By the fourth grade, I suddenly realized that I could read newspapers and books. My first joy was to find out that I was able to read the Chinese classic novel, *Journey to the West*, a mystic novel about a Tang dynasty monk who journeyed to India to gather the sutra with the help of a monkey king and a pig. This was my favorite story growing up and was my main motivation to study Chinese. I also read other classic epic novels, *The Romance of the Three Kingdoms* and *On the Water's Edge*. Such novels were considered boys' stories. Growing up with four brothers, I had heard the stories many times and was thrilled to read them on my own. I also discovered Tang dynasty poetry, which my father and mother, who studied Chinese literature at Wuhan University in Hubei during the war, recited frequently. My father was especially delighted that I liked Chinese literature and taught me some Chinese classics every weekend. He always hoped that I would become a writer or something. I was not very serious about it. In fact, I was more interested in Sherlock Holmes and Arsène Lupin. My best friend Yu-Tang Xiao in elementary school was also a fan of their adventures. She and I always traded our latest acquisitions in the form of either a novel or a comic book.

In elementary school I was also very good at arithmetic. In the last two years, we used some traditional Chinese arithmetic problems supposed to be useful for real-life applications. One of the problems is called

Chickens and Rabbits In the Same Cage: Determine how many chickens and rabbits are in the cage, given the number of heads (e.g., eight) and feet (e.g., twenty-two) in the cage.

I had no difficulty solving such a problem, but had difficulty finding it useful to me in real life. Such problems were very frustrating to kids who were not particularly mathematical, and who might even be punished for having incorrect answers! I always had great sympathy for the students who really struggled in class and were punished physically by the teacher for not being able to solve such problems. Later in junior high, I realized that with good notation in algebra, solving this kind of system of linear equations with two variables could be much easier.

In the last year of primary school, we were all busy preparing for our entrance exam. We were the last group of kids in Taiwan who had to suffer the horror of possibly failing the entrance exam at age twelve, thus ending one's schooling and hope for a future career. Both teachers and parents alike warned us that if we did not pass the exam to get into a public school, we would be sent to "tend buffalo" (figuratively speaking). I did not feel the same pressure in later entrance exams for senior high schools and colleges as I felt for this one.

Happily, I passed the entrance exam to enter the Taipei Municipal Girls Junior High School, my first choice. The following year, in 1968, free education was extended from six years to nine years, eliminating the entrance exam for twelve-year old kids. That year the average height of a twelve-year old in Taiwan shot up by one inch for boys and girls.

2.2 Girls' High School

Following the Japanese tradition of segregating girls and boys in high school, the junior high school was only for girls. But, under the Nationalist government, the curriculum and the number of public high schools were exactly the same for boys and girls. It turned out to be a very good system, especially for girls—we were really separate but equal. All students entered the public schools based on that one test, with no exceptions, so the school included kids from elementary schools all over Taipei. This was my first exposure to local Taiwanese kids, those whose parents did not come from the mainland. Their accents were curious to me. But my Sichuan accent was sometimes a source of mockery.

In junior high school, we started to learn English. Though my parents both knew English, they had never taught me the language before. When I first learned this new language, it was such an eye-opening experience for me. I could not believe that one only needed a twenty-six letter alphabet to write every word in the language. It seemed that all our elementary education was spent learning how to write Chinese characters. If one had invented an alphabet for Chinese, I could have read the Chinese classics much earlier than when I was ten years old! I learned English with all the enthusiasm of a new kid learning a new language. I also listened to records entitled "The Linguaphone" that my father bought for us to learn English. This is a language course designed by a group of linguists. There were fifty lessons altogether, and I memorized them all in about a year. Naturally my English was much better than the other students'. I was also very much into rock and roll songs, from Elvis to the Beatles to the Kingston Trio. They were all great learning tools for me.

My enthusiasm in English was short-lived. The English classes after the first year were all about English grammar, instead of teaching us to appreciate English literature or more interesting things. We were drilled every day about the plural forms of "potato" or "bamboo." I also found another course more interesting—history. At that time the history curriculum in junior high school consisted of three years of Chinese history. The history course was taught for three years by the same teacher, Ms. Shi Man-Hua. She was such a gifted teacher that it was like attending a story-telling session rather than a class. I always could not wait for my history classes.

I continued to enjoy and excel at math classes. In junior high school, the math curriculum consists of algebra, which include systems of linear equations and factorization. In the third year, we learned Euclidean geometry. I liked the simple logic of proof based on a few axioms very much. This was the first time I became fascinated by mathematics.

I got into the Taipei First Girls' High School in 1970 as expected. The school was founded in 1904 during the Japanese colonial period. During that period, the curriculum emphasized the subjects of a traditional woman's education like music, arts, cooking, and sewing. It was a famous "bride's school," since the main goal was to prepare women to be brides. After the Nationalist government took over, the new curriculum was updated to be as rigorous as at any boys' high school. During my first two years there, the president of the high school was still the legendary Ms. Jiang Xue-Chu. She dedicated her whole life to the school, which was transformed from a bridal school to one known for its academic excellence.

We were not only taught to excel academically, but also required to do well in cooking, music, arts, and athletic activities. The education also emphasized participation in all school activities, at least for the first two years. For the third-year students, there was only one thing on our minds: to pass the entrance exams in order to get into a good university. Later I learned that many girls in the U.S. did not have the experience of playing team sports in high school. This was not the case at all in our high school. Each class had to participate in every sports event, like track and field, volleyball, basketball and choir competition. Even the required daily cleaning of classrooms after classes became a competition. Since we spent so much time together, I made a lot of good friends while in high school.

In my first year, I found biology very interesting, especially when we studied genetics. By the end of the first year, each student had to choose to study science or humanities. I was sure that I wanted to study biology. At the beginning of the second year, our teacher in Chinese asked each student to write an essay on "How I Decided To Study Science." I was so sure of my choice, I wrote the article on how I would like to study biology in the future. Probably because of the clarity of the article, the piece was chosen and posted on the bulletin board for everyone to see as a model. A year later, another declaration was needed for all the science students—either biology (including medicine) or physical sciences. I chose physical sciences instead of biology. It was a constant question from other students why I changed my mind since my essay was so convincing. One reason was that, while I liked genetics, I did not like the experience of dissecting frogs or, in fact, any experiment.

In high school math classes, we used textbooks written by some college professors based on the New Math concept. It was very abstract and difficult for most students. I still remember that one semester, the math textbook started with the completeness of real numbers! During the last year in high school, my math teacher was Mr. Yang Kwan-Man. The math we learned that semester was combinatorics and probability. I was really good at it and had many perfect scores in a row. Mr. Yang always praised me in front of the whole class and constantly encouraged me to study mathematics. That was the first time that this notion crystallized in my mind. Thus when it came time for me to declare my preferences for the college entrance exams, I chose the math department of National Taiwan University as my first choice. At that time, I had not learned any calculus at all.

2.3 National Taiwan University

Like many students in Taiwan, I wanted to take a break from studying to enjoy other aspects of life after entering college. Freshman year was a new social experience for all students since almost all of us attended all-girls or all-boys high schools. It was our first time in a coed environment in six years. There were forty-five students admitted to the Math Department, and nine of them were women. Most boys had crew cuts since they had just served the required two-month military training before entering college. All the girls had curly hair since it was the first time we were allowed to wear our hair long and to have perms. Since I entered the mathematics department at National Taiwan University with the highest entrance exam score, I was made the math class representative in the first semester by default. I took this job very seriously and was busy with things like organizing excursion trips, dancing parties, etc. I had very little time for studying.

National Taiwan University was founded by the Japanese with the name "Taipei Imperial University." The campus has western-styled buildings lined with palm trees. In 1949, when the Nationalists came over to Taiwan, the president was Fu Si-Nian, the former president of Beijing University. He passed away suddenly and was succeeded by Chien Si-Liang, who greatly expanded the university and transformed it into one of the leading universities in Asia. All the students entered by merit, i.e., passing through the entrance exams, with no exceptions. The education was (almost) free. For many students, including myself, it was a dream come true. We were all so excited about our new-found freedom and status.

The curriculum at the university for each student was very rigid with not much flexibility until the junior year. In the first year, we took calculus, linear algebra and physics. Our calculus course was taught by Professor Wu-Hsiung Huang, a former student of Blaine Lawson at U. C. Berkeley. Linear algebra was taught by a teacher who was trained in the Japanese tradition, and was mostly about computation of matrices. My performance in all three courses was just above average and I almost flunked my freshman physics lab. But I did well in Chinese, English, Thoughts of Dr. Sun Yet-Sen (a political course) and the general history of China even without studying much (we all had to take seven courses each semester in our freshman year plus physical education). Many classmates were always suspicious that my high score in the entrance exam was not due to my math abilities, but because of my English and Chinese. It seemed that now they had confirmed their suspicion that, "*You girls cannot do it.*" This was publicly and enthusiastically proclaimed by some of my male classmates after the first semester at the National Taiwan University. Some of my male classmates would say such things to female students with no regard for our feelings. This negative environment was new to me. In girls' high school, our teachers were always encouraging and made us believe that we could do anything! Partly because of the insults and partly because I was bored with partying and goofing-off, I became a more serious student in my sophomore year.

The required math courses for sophomores were advanced calculus, advanced algebra and applied analysis. The textbook we used for advanced calculus was *Mathematical Analysis* by T. M. Apostol. The teacher, Professor Lung-Chi Miao, got his Ph.D. from Göttingen and had a German no-nonsense attitude in his teaching. He finished the book in one and a half semesters, covering everything from the first to the last page. Then he used the book *Calculus on Manifolds* by M. Spivak for the remaining part of the second semester. I remember that one day, in the Advanced Calculus course, professor Miao put the Heine-Borel Theorem on the blackboard:

Every open cover of a compact set has a finite subcover.

From this one theorem, I realized that this was what modern mathematics was about. Though it was not what I thought in high school at all, it was not as hard for me as it was for many other classmates. Many students transfered out of the math department after the first or second year.

For the sophomore algebra course, we used Hoffman and Kunze's *Linear Algebra* and I. N. Herstein's *Topics in Algebra*. As for applied analysis, the teacher was Professor Wei-Tseh Yang, who got his Ph.D. from Princeton. He did not use any particular book, but I learned several important topics including the Laplace equation, Green's functions and distributions in his course. He flunked almost all students except six. I was one of the six and had the highest score. After the sophomore year, my classmates had much more respect for me—so much so that they elected me to be the president of the Mathematical Society (kind of a glorified math club).

Besides mathematics courses, all the science students were required to take a second foreign language course for two years. The default language for math majors was German, but one could substitute French or Russian. Our German teacher for the math majors was an elegant professor who had a Ph.D. in comparative literature from Göttingen. He had no interest in teaching a group of unmotivated students. Though I did well in his course, I did not learn much German after one year. I decided in my junior year to move out of the assigned German course and took German from a German priest. He was a popular teacher on campus but was known to be very tough. I learned a lot more German from him but got a much lower grade. I found the grammar in German overwhelming. I never felt the same excitement with German as I did with English.

In my junior year, I took complex analysis from Professor Miao using L. Ahlfors's book *Complex Analysis* (second edition). Professor Miao finished the whole book in a little more than one semester and continued to teach Nevanlinna theory from his lecture notes in the second semester. These real and complex analysis books (pirated copies) remain my prized possessions after all the years, even though they all fell apart. I also took algebra, differential geometry and probability in my junior year. We used the first of the three volumes of *Lectures in Abstract Algebra* by Jacobson in the junior year. Though I always did well in algebra, I liked analysis much better (Fig. 2).

Figure 2. Junior Year at NTU, 1976.

In my senior year, I took three math courses: real analysis, ordinary differential equations and topics in geometry. I also took economics, but found it very boring. Most male classmates had to serve two years in the military after graduation, so they were busy preparing for the exams for being commissioned officers in the army. I decided to go to the U.S. to get an advanced degree, so I was busy preparing for the GRE (Graduate Record Exam) and TOEFL (Text of English as a Foreign Language), and sending out applications. I applied to several graduate schools—some to their math departments and some to their statistics departments. At that time, I was not sure if I could succeed as a mathematician.

The first acceptance letter came from the Princeton math department. I still remember the excitement of that day (January 28, 1977) when I received the letter offering me a full scholarship. Afterwards, I received acceptances from all the schools that I applied to, including the Harvard statistics department. Students from our math department had a good reputation abroad at that time. Being admitted to a good school was nothing new, but being admitted to Princeton or Harvard still caused some excitement. Some teachers said that I should go to Princeton, but others advised that it was easier to succeed in statistics.

I asked my mother. She did not know that I had applied to Statistics Department also and scolded, "If you want to study mathematics, study mathematics, why bother with statistics?"

In her opinion, pure math is the higher form of scholarship, while statistics is for people who buy and sell. She always looked down upon business people. I told her that I worried that I might not succeed in mathematics. My mother said,

> The fortune teller said that you are smart and strong, better than men. You are always strong when the opponents are strong, weak when the opponents are weak.

Sometimes a mother's blind faith is all one needs! My mind was made up. The day I accepted the admission offer from Princeton, I also took out my book *Real Analysis* by H. L. Royden and studied it in earnest. I said to myself, "The game is afoot."

3 Becoming a Mathematician

3.1 Princeton, 1977–1981

I arrived at Princeton in July 1977. After staying for a few days with my sister, who lived nearby, I moved into the Graduate College. The castle-like dormitory for graduate students was more beautiful than in the picture on the cover of the graduate school brochure. It bordered a golf course and had a tall tower, the Cleveland Tower. I enrolled in an English language course for foreigners on campus. The students in that course came from all over the world. My initial culture shock was mollified by the beautiful surroundings among the Gothic buildings and tall ancient trees. Every day I carried the bookbag that my college classmates gave me as a going away present. After two summer months, my spoken English improved.

In September, the semester started. To my surprise, I found that I was the only woman student out of eleven in my class, plus two male visiting students from Germany. Not only that, but there was only one other woman student, who had transfered from Chemistry, in the whole graduate math program. It seemed that the math department had not admitted too many women graduate students before me. There were no women professors on the faculty either, only one woman instructor from Germany. The first year I arrived at Princeton, these were the only women mathematicians at the university, in addition to a few visitors that I met at the Institute for Advanced Study. When I was young, my father always liked to tell me stories about women scientists. One was Madame Chien-Shiung Wu, the famous physicist who became the first female instructor at Princeton and later the first female professor of physics at Columbia. The other was Tsai-Ying Cheng, a graduate of National Taiwan University who became the first woman to receive a degree (Ph.D. in biology) from Princeton in 1964, five years before the university started admitting undergraduate women students. It was only in my second year at Princeton that the math department hired Chuu-Lian Terng as the first woman assistant professor. Chuu-Lian also graduated from National Taiwan University and went to the same high schools as I did, only six years before me. We immediately developed a special camaraderie and became close friends. Three decades later, another National Taiwan University graduate, Sun-Yung Alice Chang became the first woman department chair at the Princeton math department (Fig. 3).

1. Mei Chi Shaw
2. Greg W. Anderson
3. Brian C. White
4. Robert F. Coleman
5. Dieter K. Bassendowski
6. Mark I. Heiligman
7. John P. Snively
8. Eric R. Jablow
9. Thomas G. Goodwillie
10. Roderick D. Ball
11. Wolrad B. Vogell
12. Don M. Blasius
13. Allan T. Greenleaf

Department of Mathematics

Entering Graduate Students - Fall 1977

Figure 3. Princeton 1977 Entering Graduate Students.

The Princeton math department was unique in that it did not offer any basic graduate courses. The idea was to throw all the graduate students into the most advanced courses right from the beginning, a kind of total immersion from the start. There were no required courses, no grades. The only exam a Ph.D. student needed to pass was the General Exam, within the first two years. It was a three-hour oral exam on two advanced topics chosen by the candidate and three basic courses: real and complex analysis and algebra. I knew that I wanted to study analysis, but I had no idea which subject in analysis. I decided that I would sit in all three analysis courses offered that semester by Professors R. C. Gunning, J. J. Kohn, and E. M. Stein. Though I had been forewarned by my teachers at National Taiwan University about the advanced nature of the courses, I did not know how ill-prepared I was for this new system of learning. After only one week, I stopped attending Professor Kohn's course since I was totally clueless. Another two weeks later, my anxiety had reached such a level that I really did not know if I should continue. At that time, the title of Stein's course was "Another Class of Pseudo-Differential Operators." There were more than twenty people in the audience. I knew a little bit of differential equations, mainly the Laplace equation. I had not heard of pseudo-differential operators, let alone "another class" of them. I mustered all my courage to see Professor Stein. I bluntly asked him two questions:

(1) should continue to attend his course, since it was way over my head?
(2) should study mathematics at all?

Professor Stein told me that he did not expect me or any first-year graduate student to understand much in his course.

"Just look at the audience in the class," he said. Half of the attendees were professors, and the others included advanced graduate students. He suggested that I should still sit in the course and hopefully, in a year or two, I would understand more.

As for the second question, he said that since I was already there, I might as well stick around and see what happened. He also suggested that I read his two books, *Introduction to Fourier Analysis on Euclidean Spaces* (joint with Guido L. Weiss) and *Singular Integrals and Differentiability Properties of Functions*. I had brought both books (pirated copies) with me from Taiwan. I studied the two books thoroughly and they somehow calmed me down. I still attended the courses but did not worry about whether or not I understood anything.

I decided to choose Fourier analysis and several complex variables as the two topics for the General Exam. Since I was attending Gunning's course, I was reading the book *Analytic Functions of Several Complex Variables* by R. C. Gunning and H. Rossi to prepare for the exam. At the same time, I was also reviewing the three basic topics. Fortunately, my undergraduate training in analysis and algebra was quite solid. But before taking the General Exam, one has to pass foreign language requirements. Each student had to demonstrate the ability to read math texts in two out of three of German, French and Russian. My two years of German in college came in handy, so I passed the German, examined by Professor Gunning, without much difficulty. However, I really did not have time to learn French. Other graduate students told me that the professor always asked students to translate the same book, *Introduction à L'Étude des Variétés Kählériennes* by André Weil, so I simply memorized the English translation of the first chapter of the book. When I went to take the exam with Professor Moore, however, he gave me a French calculus book instead and asked me to translate the first few pages. He thought that it would be easier for me. I read the first page, which started with "Nombres," or "Numbers." Though I could roughly guess the meaning, my translation was very halting and probably incorrect. Finally I told him I had studied the book by André Weil but not this book. He gave me Weil's book and asked me to translate the first chapter. That is how I passed the French exam. I passed the General Exam in May by the end of the first year. Thus I survived the first year at Princeton after a shaky start.

After the exam, I decided to work with Professor Kohn in several complex variables since I always liked complex analysis. Professor Kohn was the first one to solve the $\bar{\partial}$-Neumann problem on strongly pseudoconvex domains, and his solution is called "Kohn's solution." The Gunning and Rossi book on several complex variables, which I had studied for my exam, uses sheaf theory, a more algebraic approach. It is completely different from Kohn's approach, which uses partial differential equations. The first thing for me to do was to read Folland and Kohn's book, *The Neumann Problem for the Cauchy-Riemann Complex* and to read Kohn's original papers. After about one year, the Folland-Kohn book had fallen apart and I gradually understood the subject. In my second year, I sat in the same courses under the same titles as the first year, but they began to make sense to me now, when in the first year, they had all seemed intractable. Kohn lectured on his influential paper on sufficient conditions for the $\bar{\partial}$-Neumann problem using the multiplier ideals, a paper he had just completed.

After one semester in the second year, Professor Kohn gave me a warm-up problem to do before giving me the real thesis problem. The warm-up problem was to study the $\bar{\partial}$-Neumann problem on piecewise smooth strongly pseudoconvex domains. I remembered distinctly that he told me that he expected me to finish the warm-up problem in three to six months. He also emphasized that it would not be enough to be a thesis problem. The real thesis problem will be on multiplier ideals. At that time, the solution of the Cauchy-Riemann equations on piecewise smooth strongly pseudoconvex domains had already been obtained a few years prior using integral kernels by Range and Siu. Thus he thought it would be a routine exercise to do the $\bar{\partial}$-Neumann problem on such domains. He asked me to do a report on the Range-Siu paper on the solutions for the Cauchy-Riemann equations using kernel method. It was very difficult for me to read the paper without prior knowledge of the kernel method. I muddled through the paper and did a report in Kohn's graduate course, but it was years before I completely understood the kernel method in several complex variables. At the same time, I was also studying Hörmander's *An Introduction to Complex Analysis in Several Complex Variables*. Kohn's subelliptic estimates, Hörmander's L^2 method, and the kernel approach were the three major attacks on the Cauchy-Riemann equations. Later I realized that perhaps the idea of total immersion might not be too bad. By exposing graduate students early to the forefront of research, they had a broader scope and became very independent (for those who survived). At that time, I felt very inadequate, and I was not alone.

During my third year, there were even more activities in the department and at the Institute for Advanced Study in Princeton. Charles L. Fefferman was giving a course in the department for the first time after many years on leave. I was taking courses from John Erik Fornaess and learned a lot of the counterexamples in several complex variables. I also took courses at the Institute for Advanced Study that year, including lectures by Louis Nirenberg and Shing-Tung Yau on nonlinear partial differential equations. That was a special year in differential geometry at the Institute, organized by Shing-Tung Yau. Every Wednesday morning at 8:00 am, Yau would give a lecture on nonlinear equations and his solution to the Calabi conjecture. The auditorium room was packed with mathematicians from all over the world. I still remember the excitement one felt during those lectures. I also met many mathematicians that year, including Professor Shing-Shen Chern. At Princeton, one was always surrounded by many brilliant people, not only people in several complex variables, but also in other fields. At times it could be intimidating, but it was most stimulating. One thing I learned early there at Princeton was that one has to work very very hard, and hard work is only the necessary condition. I still do not know what the sufficient conditions are, if there are any.

After working on the warm-up problem for almost one year with not much progress, I was very frustrated. Everything turned out to be the opposite of what Kohn had expected on non-smooth domains. The boundary value problem on smooth domains cannot be generalized to non-smooth domains except the basic L^2 theory, which had been obtained in Hörmander's seminal paper in 1965. Anything beyond that required new techniques, so it was premature to expect to solve the problem at that time. Even the easier Hodge theorem on non-smooth (Lipschitz) domains was not understood at all. I obtained some results on the special case for the Hodge theorem on Lipschitz domains with conical singularities. That turned out to be my thesis. The Hodge theorem on general Lipschitz domains was not solved until three decades later, published in 2001 by Dorina Mitrea, Marius Mitrea and Michael Taylor in a Memoir of AMS (American Mathematical Society). A few years later, Dorina, Marius and myself gave another proof of the theorem. At the time, I felt very bad about not being able to solve even the warm-up problem.

3.2 The Winding Road to Notre Dame

After I finished my Ph.D. at Princeton in the summer of 1981, I took a postdoc position at Purdue University. The head of the Math Department was Salah Baouendi. I had met him the previous year at a conference organized by F. Treves at Rutgers. He was an extremely energetic person and very active in research. There were a lot of activities and people around him at Purdue. When I first arrived at Purdue, I only knew that I did not want to continue my graduate work on domains with corners. I attended the seminars and talked to Baouendi and Baouendi's student, Chin-Huei Chang, often. Chin-Huei Chang was also from Taiwan University and he was working with Baouendi and Treves on the real-analytic hypoellipticity of vector fields. Within a few months, I found a problem related to their question, but in the smooth category. The special case of this question was already known by the Kohn-Rossi theory, so it was related to what I had studied before. Within a few months, I solved the problem.

Right after I finished writing the paper, a former student of Hörmander and a visitor at Purdue that year, Anders Melin, told me that the result I got was already known earlier from Egorov-Hörmander's theorem, once the problem is micro-localized. I was devastated by this blow. Baouendi told me that I should still submit the paper since my new method itself would justify a paper, but make it clear that the result was known from the microlocal technique. I submitted the paper and it was accepted later that year. He also gave me very practical advice on how to write a paper or a proposal. He asked me to submit a proposal to NSF for my own research, even though the chances of getting funded were small, and he even gave me a copy of his own proposal as an example to follow. He advised me to read the classic papers by Kohn and Hörmander and imitate how they write. I remember his sometimes blunt advice clearly: "Every sentence has to have a comma or a period. Do not invent your own English." Later, I often repeated this same practical advice to my own students and postdocs.

At the beginning of the second year, I had to look for another job since my position at Purdue was only for two years. Having only two papers accepted but not yet published (the other one was based on my thesis), I did not hear anything from any department. By the end of March, I was about to be ready to make a "Plan B" when a phone call came from the head of Texas A&M, Dr. E. Lacey, inviting me to interview for a tenure-track position. At that time, Al Boggess was the only person I knew in the department and he was the main advocate for my hiring. I got the job in April and no Plan B was necessary.

That summer, I married my husband, Hsueh-Chia Chang, a fellow graduate student from Princeton. He was born in Taiwan, but left when he was eight and lived in Malaysia and Singapore before coming to the U.S. But he was teaching at UC Santa Barbara at that time. The two-body problem was not easy to solve either.

Texas A&M is in the middle of nowhere, just like Purdue, but everything is bigger in Texas. The school was also unique because of the presence of so many cadets on campus. They always addressed me as "Ma'am." It took me some time to get used to the Texas drawl. But mathematically, we had very stimulating seminars with many young participants who had similar interests. One year later after I arrived, Harold Boas was also hired.

Al Boggess was a student of John Polking and he was an expert on the kernel approach in complex analysis. In the seminars, he gave a series of lectures on integral kernels and was a great teacher. It was there I learned the kernel approach to the Cauchy-Riemann equations and the tangential Cauchy-Riemann equations. The paper by Range-Siu that I studied by myself with great difficulty as a graduate student now became easy to understand with the new interpretation using the work of Harvey-Polking. Very soon Boggess and I found a problem to work on together and wrote a paper on the local solutions to the tangential Cauchy-Riemann equations. This was the first collaboration I had with another mathematician. I realized the importance of having collaborators. Not only did each of us bring new ideas and different perspectives, but the interchange of ideas was so stimulating that new results could come out that would otherwise be impossible.

I also continued my research on the global L^2 existence and estimates of solutions for the tangential Cauchy-Riemann equations. This work started when I was at Purdue. I continued to expound the problem and wrote two papers on related subjects. But there was one major obstacle to proving the main theorem. One day after teaching a course on finite mathematics (high school math), a simple idea just came to me: using change of variables in calculus. Once this was thought through, I wrote the paper, L^2 estimates and existence theorems for the tangential Cauchy-Riemann complex. I decided to submit the paper to *Inventiones Mathematicae*, one of the best journals. I waited anxiously for five months until a letter arrived from the editor-in-chief informing me that the paper would be accepted upon satisfactory revision. There was still one case missing. I discussed the problem with Boas who was working on the regularity of the Szegö projection at that time. We solved the remaining case in a few months and submit the paper to *Mathematische Annalen*. These papers established the L^2 theory and closed range property for the tangential Cauchy-Riemann equations on pseudoconvex boundaries. I also wrote a paper on non-linear partial differential equations as a result of talking to H. Brezis while he was visiting there.

After two and half years at Texas A&M, I decided to join my husband at the University of Houston, where he took a job as an associate professor. Without people of similar interests in the department, I realized how much I missed the collaboration I had with people at A&M. At Houston, I tried to attend the seminars at Rice University as often as I could, but it was not the same. After the department botched my early promotion to associate professor, I felt underappreciated there. I called Salah Baouendi to tell him about the situation. He told me to calm down since it was only an early promotion. He added that he would see what he could do. A few days later, Nancy Stanton, a math professor at the University of Notre Dame whom I had met at Princeton, called to ask if I might be interested at a position at Notre Dame. I went to Notre Dame for my interview on January 28, 1987 (exactly ten years after the letter of admission from Princeton), one of the coldest days I can remember. After the interview, Pit-Mann Wong invited Wilhelm Stoll and me to his house for dessert. Pit-Mann grew up in Hong Kong, but he attended National Taiwan University and was a classmate of Chuu-Lian Terng. Two weeks later, I received a phone call from the chairman, Bill Dwyer, offering me a tenured associate professorship at Notre Dame. I had been admitted to a very exclusive club of tenured professors of mathematics at a major research university. The criterion for membership includes proving a few good theorems. One month after I got my job at Notre Dame, my husband also got a full professorship at Notre Dame. The two body problem had now been solved. I was promoted to full professor in 1992.

After tenure, the rules of the game became different. The new struggle now was not to survive, but to become a better mathematician and to help others with similar aspirations. I was always flattered when people considered me to be a role model. I consider it a great privilege to have the opportunity to work with many mathematicians. I also received numerous invitations to give lectures or talks from all over the world, including countries in Asia and Europe, plus Egypt and Morocco. These invitations were great experiences for me. I worked on problems that I considered important and strived to be the best that I could be (Fig. 4).

Though I decided not to continue to work on the problems on domains with corners after Princeton, these problems were always on my mind. Since I came to Notre Dame, tremendous progress was made on the Dirichlet and Neumann problems on Lipschitz domains. Harmonic analysis on Lipschitz domains became a central theme of modern research. This made solving the $\bar{\partial}$-Neumann problem on Lipschitz domains possible. Finally I was able to prove, in a joint work with Joachim Michel, some partial results for the $\bar{\partial}$-Neumann problem on piecewise smooth or Lipschitz strongly pseudoconvex domains. I gave a talk at a workshop in November of 1995 at the Mathematical Sciences Research Institute (MSRI), founded by Chern in Berkeley, California. After my talk, some people in the audience who knew me from Princeton joked, "Finally she was able to do something with her thesis." They did not know that it was only the warm-up problem for my thesis. I also began to branch out into various different areas, including problems in complex geometry, on which I worked with Jianguo Cao for more than ten years until his untimely death.

Figure 4. Professor Kohn's 60th birthday conference, Princeton, 1992. (Picture taken by Dan Burns.) Front row (from left to right): John Stalker, Gerald Folland, John D'Angelo, Joseph J. Kohn, Donald Spencer, David Catlin. Second row: So-Chin Chen, Pengfei Guan, Mei-Chi Shaw, Lop-Hing Ho, Ricky Diaz. Third row: Jeff McNeal.

One of the most rewarding experiences as a mathematician is to teach students, both graduate and undergraduate. I take great pride in seeing young graduate students or postdocs mature into mathematicians. Based on years of teaching experience to graduate students on several complex variables, I decided to write a book on the subject with So-Chin Chen, another former student of J. J. Kohn. We met in college when he was a major in electrical engineering but was taking the same complex analysis course as I was. We had similar outlooks on the subject and the two of us complemented each other. Our goal with this book was to bring the subject up to date and to write it as clearly as possible. We joked that it must be made so clear that "*we could explain it to someone on the street.*" During the next four years, writing the book became so all-consuming that I was always glad when it was over. The book *Partial Differential Equations in Several Complex Variables* was published in 2001 and was well-received. It has been most rewarding to hear from people, from the most prominent researchers to beginning students, who find the book useful.

My proudest moment was when Shing-Shen Chern wrote the preface of an article "On Several Chinese Women Mathematicians" in the Chinese magazine *Zhuan Ji Wen Xue* ("Biographical Literature") in 1995[1]. This is still a popular magazine in Taiwan. Chern's own autobiography, "Studying Mathematics for Forty Years," appeared in that magazine in 1964. In the preface, Chern called the six women mathematicians from National Taiwan University, Sun-Yung Alice Chang, Wen-Ching Winnie Li, Fan Chung, Jang-Mei Wu,

---

[1] For a copy of the article, see http://www3.nd.edu/ meichi/miscellaneous.html

Chuu-Lian Terng and myself, "... a miracle in Chinese history; the glory of the Chinese people." The first four were in the same National Taiwan University class in 1970 and became legendary in Taiwan. I did not bother to tell my family in advance, so when my father read the article, he was beside himself, as if I had just won the Nobel Prize. When I assured him that I was not that famous nor did I deserve such praise, my father said, "If Mr. Shing-Shen Chern said so, then you are to me." (Fig. 5).

Figure 5. Woman Mathematician Forum, National Taiwan University, July, 2009. (http://www.tims.ntu.edu.tw/exlink/ntumath2009/week2-2.html.) Panelists from left to right: Yng-Ing Lee, Wen-Chin Winnie Li, Sun-Yung Alice Chang, Chuu-Lian Terng, Mei-Chi Shaw and Fan Chung Graham.

I had met Chern a few more times after I had left Princeton. Over the years, I always reminded myself, "You must not make the great mathematician Chern look like a fool." In 2001, I went to a special conference in honor of his ninetieth birthday in Taipei hoping to thank him in person. Unfortunately he was too weak to make the trip from Tianjin, China. I wrote a letter to thank him again for his extreme generosity in praising Chinese women mathematicians, along with a copy of our book. He graciously acknowledged the receipt of our book (with more praise) and invited me to visit him in Nankai, Tianjin. I wished I had.

Looking back, I have been very lucky to have had jobs at Purdue and Texas A&M, which later became powerhouses in several complex variables. The two places might not have been the most exciting places to live, or the most glamorous. But for me, they were the best places for me at that stage of my career. It was a time for me to slowly mature into a mathematician. Notre Dame is another place in the middle of nowhere, but it suits my disposition well. Along the way, I have been helped mathematically and otherwise by many friends and colleagues. I had also been extremely lucky to have been able to do the things I love for a living. It was sad for me that Wilhelm Stoll, Pit-Mann Wong, Jianguo Cao, and Salah Baouendi all passed away within the past three years. I will always remember their warm friendship in my heart (Fig. 6).

Figure 6. Giving a talk at Phong's 60th birthday conference, Columbia University, May, 2013 with Louis Nirenberg and Xiaojun Huang in the audience. (Picture taken by Christina Sormani.)

4 Journey to the Village Leiyang, Hunan

In May 2008, I visited the Math Department at Fudan University in Shanghai for one month. Since my first trip to China in 1987, I have visited China often, mostly to give lectures at universities. This time, my younger brother Chang-Yu was also in Shanghai on a business trip. One day I called him to ask if he was interested in visiting Hunan Leiyang, our mother's hometown. He said that he was also thinking about it. Thus the two of us embarked on a root-finding trip to Leiyang years after both our parents passed away.

We first flew to Changsha, the capital of Hunan, and took a bus to Henyang, the largest city near Leiyang, where my parents took the last available flight out of Hunan to Taiwan in 1949. After staying at a hotel in Henyang, we met our cousin, my mother's nephew, the next day before we hired a taxi to the Leiyang village. The four-lane highway became two-lane, then one-lane before it turned into a dirt road across the rice fields. The taxi driver refused to continue for fear that the car might fall into a ditch. He agreed to continue only after we promised to pay him more. After a few turns on the dirt road, suddenly a small village appeared in the distance before us. The village was hugged by small hills at the back. It was such a beautiful sight that it reminded me of those small Italian towns in Tuscany, but surrounded instead by lush green fertile filelds. The taxi driver exclaimed, "This is really good Feng-Shui!" Even he was excited.

After we passed a pond, we first saw a very symmetrical two-story building with a Chinese character "Shou" (meaning longevity) written in the middle of the second story. Our cousin told us that it was the ancestral worship place of the Liu family. The building was built of local blue bricks, famous for their durability. Leiyang is not only rich in agricultural products, but also rich in a special kind of coal. The coal generates an especially high temperature in order to make the blue bricks. The brick is actually white but develops a bluish patina after years of wear, which makes it more beautiful. The building was now empty with nothing left in the interior and all the windows were gone, yet it was still standing with its beautiful exterior.

While I was admiring the beautiful remains of the old building, an old gentleman came and asked whom I was looking for. It was unusual for strangers to come to the remote village. I told him that I was not looking for anybody, but my mother used to live here. He looked surprised and asked who my mother was. I told him that my mother's name was Liu Lan-Fang. He looked more than surprised, almost shocked. Then he told us to wait for a moment. Within a few minutes, the old gentleman came back with another old man, to whom he introduced us. The old man did not say a word, but took us to the front door of my mother's ancestral home. The door was gone but the two large columns that framed the door were still there. My mother used to tell us that her family motto was inscribed on the columns by the door: "Art and Literature enlighten the world. Loyalty and Kindness run in the family." One could still see some characters on the column. There used to be two lions in front of the gate of each house, but now the lions were gone, destroyed during the Cultural Revolution. When we entered the gate, there was a courtyard in front of us with two-story brick buildings on both sides. I could only imagine what a beautiful country home it must have been, but now it was very dilapidated.

The old house had more than thirty rooms and was now occupied by eight families. The old man pointed to a room where my mother used to live, but now the room was locked and we could not enter. We continued across a low gate and came to the kitchen, the wine cellar, and the stable. When we were young, we used to laugh at our uncle's stories about horse riding and hunting as jokes. We always thought that our uncle was bragging. Now it turned out to all be true. Further down, after exiting the house, we entered another house similar to the previous one. That house belonged to my mother's uncle. The two brothers lived side by side just like in the classic Chinese novel, *Dreams in the Red Chamber*. My grandfather died when my mother was only eight. Her uncle was like a surrogate father to her.

By the time we finished touring the house, all the villagers had come to see the "two strangers." It was a scene that reminded me of a story my father told us. When he came to my mother's home to meet her family for the first time at the end of the war, the whole village came out to see "Lan-Fang's husband" before he even entered the house. My mother's family was very much against their marriage in the beginning. My parents fell in love and decided to get married in Chengdu, Sichuan during the war, despite opposition from both families. Such a thing was unheard of at that time in the village. Her family had other plans for her marriage, but after a beautiful grandson was born, my mother brought the child back to her family and they finally reconciled. To the great sadness of my parents, the child died of strep throat at age three in Leiyang. When my father visited there, the family welcomed the handsome son-in-law wholeheartedly with a feast that lasted three days.

I suddenly realized why my mother was very often unhappy when we were young. Growing up in Taiwan, there did not seem to be a day that our mother would not miss her home and the people in Leiyang. The sudden change from the richest family in Leiyang to the almost refugee-like existence in Taipei was probably too much for anyone to take. When my mother left home in 1949, she was only thirty. By the time she returned, she was sixty-seven. Seeing how the home lay in ruins, she was heartbroken. Yet she was an optimist, and did her best to raise six kids under the most trying circumstances.

On our way back to Shanghai, my brother and I sat in total silence. We always knew that our mother was special. Now we knew just how special and why (Fig. 7).

Figure 7. My mother's ancestral home in Leiyang, Hunan.

5 Journey to Sichuan

My mother was not the first woman from her village to attend Wuhan University—she was the second. The first was her cousin, Liu Lang, who was considered a revolutionary in her time. She defied her family's arranged marriage and later married a communist. My mother admired her very much. After attending the elementary school run by her family, she followed the footsteps of Liu Lang to attend high school in Changsha and then Wuhan University in 1938. At that time, the Sino-Japanese war had already been going on for more than a year. The Nationalist government had lost Shanghai, Nanjing (the capital city), and the eastern half of China, and had reestablished the government at Wuhan. Within a few months, the government decided to move further west to Sichuan, a province with fertile land surrounded by tall mountains. It is a place called "Heaven on Earth" with natural barriers, easy to defend and difficult to invade. The administration at Wuhan University also decided to move the entire faculty and students to Leshang, Sichuan.

The journey from Wuhan to Sichuan was most treacherous. By boat, one has to pass through the Three Gorges, whose famously beautiful scenery has inspired many poets and writers over thousands of years. There was no highway at that time. By train, one could only get to Quiyang, Quizhou, and go from there. But the tickets were difficult to get since all the available resources were used to transport government and military personnel. My mother and her best college girlfriend got on the train from Wuhan to Quiyang. It was on this train that she first met my father. As fate would have it, my mother and her friend boarded the same compartment where my father and his fellow officers were sitting. He was a dashing twenty-six year old Air Force officer dressed in full uniform. My mother was not yet twenty, dressed in her westernized Chinese clothes. Immediately they struck up a conversation and found out not only that they both loved literature, but that they shared similar dialects. Though my father was from Guanxi province, a province adjacent to Hunan, the village he grew up in was very close to Leiyang. There is such a thing as love at first sight.

I always wanted to visit Sichuan to retrace the footsteps of my parents in Chengdu and Leishan. For some reason, I never got to do it until 2012. The moment I got off the plane in Chengdu, the capital of Sichuan, I thought I was back to the Taipei Air Force Primary School again. The language came back to me, though I had had little chance to use it except when my best friend Yu-Tang came to visit me. Not only the language, but the food and the people seemed all familiar to me. The Sichuan people have a reputation of being hot-tempered, which many attribute to the hot spicy food. The Sichuan women were noisy and laughed out loud, just like us. I realized I have a lot of Sichuanness in me. I also visited Leishan, where my mother studied for two more years before marrying my father. Leishan is in the foothills of Emeishan, the sacred buddhist mountain. It is also at the confluence of three rivers. There is not much left in Leishan today except the famous Giant Buddha, still guarding the rivers.

One of my favorite pictures of my parents was taken in 1939 in Chengdu at the tea house in "DuFu Thatched Cottage." DuFu was my mother's most favorite Tang dynasty poet. My father took the picture of the two of them and gave it to my mother. The two of them looked so handsome and so happy together during that period in the middle of the Sino-Japanese war. I have the picture on my desk in my home office in Indiana. Last summer, when I sat at exactly the same place in Chengdu, I realized that perhaps Sichuan was the place that they wanted to remember. This fond memory existed in many people in the Village of Righteousness and survived to the next generation in me and my Sichuanese-speaking classmates. Most parents in that village spent many years in Sichuan fighting the difficult Sino-Japanese war, but it was a war worth fighting. Later, after losing the civil war and the hasty retreat to Taiwan, they tried to cling to the good memories and forget about the painful defeat. They also did their best to cling to the Chinese traditions and teach their children to be proud even under most difficult conditions. Thus every Chinese holiday was celebrated with fanfare and the Chinese culture was proudly preserved on that part of the remote island. Over the years, I always felt that I cannot fail, since my failure would be not just for myself, but for two generations. I carried the torch of their dreams from Sichuan. I can only hope that I have lived up to the premonition of the fortune teller, which my mother so firmly believed (Fig. 8).

Figure 8. My Parents in Chengdu Sichuan 1939.

Acknowledgements

I would like to thank professor Steven Krantz for inviting me to write this article and Professor Ming-Chang Kang for his comments on the curriculum at the National Taiwan University. I would especially like to thank my niece Cathy Wu for her editing of the first draft, which has greatly improved the English and style.

Part III

Why I Became a Mathematician

Foreword to
Why I Became a Mathematician

Many of us who are mathematicians do not spend a lot of time fussing over *why* we became mathematicians. It was something that we evolved into, and it happened naturally.

But an invitation to write for this portion of our book gave several active mathematicians an opportunity to really dwell on this question. And the results are informative and inspiring.

Some people became mathematicians because they love to solve problems, others because they love to teach or communicate, others because they love to develop theories, still others because they love logic. In many, if not most, instances there was a key teacher or mentor who set an example and *inspired* the individual in question to pursue the mathematical life.

And this is an observation not to be downplayed. It would be virtually impossible for a person in complete isolation to develop into a mathematician. There must be some teaching or mentoring or instruction involved. Even the great Ramanujan noted a particular text that he found inspiring.

It is always instructive to pause and endeavor to determine how and why we became who we became. Such rumination helps us to develop a sense of ourselves, and also a sense of what mathematics is, what it means to us, and how it has played a role in our lives. It is unlikely that a carpenter would engage in similar cogitations. There is something special about mathematics that inspires us, and gives our lives meaning. One of the main purposes of this book is explicate this point, and to make it understandable to non-mathematicians.

Perhaps reading these words will attract some new people to consider taking up mathematics. That would be a happy outcome for all concerned.

— Peter Casazza, Columbia, Missouri
—Steven G. Krantz, St. Louis, Missouri
— Randi D. Ruden, University City, Missouri

20

Why I Became a Mathematician: A Personal Account

Harold P. Boas

I became a mathematician because of my kindergarten teacher.

That claim may surprise you if you know something of my family history. My mother, Mary L. Boas, was a professor of physics at DePaul University in Chicago. My father, Ralph P. Boas, was a professor of mathematics at Northwestern University and a prominent researcher, teacher, writer, editor, and translator; he was active in the affairs of both the American Mathematical Society and the Mathematical Association of America. With such parents, had I not a predetermined future? I think not, for my two siblings showed no inclination to pursue a scientific or an academic career.

Certainly all three of us children were profoundly influenced by our parents' passion for teaching and learning. We absorbed by osmosis the conviction that every craft can be mastered through diligence. For instance, we observed that when our parents took up photography, they went so far as to convert the utility closet into a home darkroom to develop their own film; their gardening activities expanded to include forcing bulbs under lights in the basement, building window boxes, and constructing cold frames in the backyard; they taught themselves how to work with fiberglass in order to maintain a succession of small sailboats. My father took pride in being able to shingle a barn, to fell a tree, and to miter a joint; my mother knew how to sew her own clothes, to make beach-plum jelly, and to wring a chicken's neck.

As the youngest child, I felt overshadowed by my accomplished siblings. Both of them mastered the unicycle and the trampoline; in sailing camp during the summer, they learned the ropes well enough to participate in regional regattas; my sister took up horseback riding and brought home ribbons from competitions; my brother learned both slalom and trick skiing and participated in water-skiing meets. Being younger and smaller, I could not match my siblings in athletic pursuits, so I turned to more cerebral activities—especially chess, in which I was nationally ranked in my age group during high school.

Now there are many attractive intellectual endeavors besides mathematics, so what happened in kindergarten that set the course of my life? Before reaching school age, I picked up knowledge at home in the haphazard way of young children. The house was full of books, and I was beginning to read when I went off to school. But my kindergarten teacher figuratively slammed the book shut. She had the crazy notion that kindergarten children should be protected from the stress of such onerous first-grade burdens as reading. She had no such strictures about numbers, however, so I redirected my energy. My parents fondly recalled that I would bounce on their bed in the morning demanding arithmetic problems to solve. One of my earliest memories is sitting in a corner painstakingly working out 12×12 by the multiplication algorithm.

The upshot was that I was always a year ahead of my schoolmates in math class. During twelfth grade, having already finished the high-school calculus course, I got special dispensation from the school administration to work through my mother's *Mathematical Methods in the Physical Sciences* as an independent-study course.

But there is more to the story, for at age eighteen I scarcely knew what mathematics really is. Moreover, I had broad interests, as evidenced by the five advanced-placement exams that gave me sophomore standing as I started college. With a push in the right direction, I might have become a specialist in novels of the Regency era (one of my prized possessions is the working copy of *Pride and Prejudice* from which my paternal grandfather taught); a historian of Renaissance Florence (did you know that 1492 is the year of Lorenzo de' Medici's death?); or a researcher in linguistics (my high-school English class studied the rudiments of Chomsky's transformational grammar).

It was a close call, for in my first year at Harvard, mistakenly thinking that I was a hotshot, I enrolled in the legendary Math 55, which at the time used Dieudonné's *Foundations of Real Analysis* as a textbook. Woefully underprepared for the course, I got discouraged with mathematics for a few semesters and turned my attention to physics. I even took a year of Russian, intending to learn enough to read the leading chess periodicals. Then, in my third year, I had the good fortune to enroll in a course in applied mathematics for which Dick Gross was the enthusiastic, encouraging, and stimulating teaching assistant. This experience was my mathematical epiphany. Determined to go further in my fourth year, I signed up for complex analysis from Lars Ahlfors and algebra from David Mumford and John Tate. After sitting at the feet of these gurus for a year, I was a lifelong convert to the religion of mathematics.

21

Why I Became a Mathematician?

Aline Bonami

Becoming a mathematician requires many ingredients. One of them is to enter the academic world (at least in France). This was the most natural thing for me to do, which is demonstrated by the fact that my two brothers also are academics. Indeed, two of my great-grandfathers were teachers during the second half of the nineteenth century. My grandmother (on my father's side), then both my parents, were teachers in high school. It would have taken a lot of energy to do something different (I wanted to be an engineer at one time), at least for me: my mother dreamed of having her elder son becoming a diplomat, but she had no such dream for me, except that I be successful in my studies. My parents were lucky on this point: my brothers entered the prestigious Ecole Normale Supérieure and in 1963, I entered its variant for girls at the time (the so called Ecole de Sèvres), where my grandmother had studied mathematics seventy years before me.

Choosing a topic was not such a natural familial choice: my grandmother and my mother taught mathematics, but my father taught humanities. My eldest brother, Georges Nivat, chose to study Russian and became a well-known slavist, while the second one, Maurice, started studying mathematics, then became a well-known computer scientist. Even if he was seven years older than I, Maurice and I spent a lot of time together before he left home as a student. But, apart from special occasions we did not speak of mathematics: he was the leader and we cycled, climbed trees, he took me for rides in the forest, we played cards, collected stamps, and shared the same passion for medieval art (it was in the fifties, we had of course no TV and only a few toys, even if we were lucky children). He taught me a lot of things, among which history was probably the most prominent. One year we decided to speak with a secret alphabet. Just A = 1, B = 2, etc. We spoke rapidly this way during one week, after which the cousin who was with us started to understand a little of what we were saying. We then decided to take the alphabet in reverse.

Maurice and my mother got interested in my training in mathematics when I was 10. At that time one had to pass a first exam to enter secondary school. To get me prepared for the examination in mathematics, my mother found a book that was full of tricky problems with trains that meet or tanks that are emptied. It happened that I found some answers while they felt unable to find them without the use of algebra, so that they could not give me explanations. This was quite exciting.

I felt much more excited on the second occasion, when Maurice taught me mathematics. He was at that time graduating in mathematics and followed a course of Gustave Choquet on Modern Mathematics, which was completely new in 1957 at the University of Paris and attracted much interest from students and colleagues. On three or four Sundays he repeated for me the course he had just listened to during the week on set theory, with quotient sets and cardinality. It was fascinating to be able to understand what appeared to me as the foundations of mathematics, much more than the elementary algebraic computations that we studied in school.

The influence of my mother and Maurice was counterbalanced by the intellectual influence of my father, which was certainly the most important that I received from my family. Listening to him speaking of literature or history was a great pleasure, which was not interrupted when I turned to scientific studies. I rapidly realized that it was a kind a treasure that would follow me in any case. Also his explanations were always very clear and simple, which is what one looks for in mathematics.

Family environment is not sufficient by itself, and teachers certainly play a central role in the development of intellectual choices. As far as secondary school is involved, I do not think that the personality of my teachers has been so crucial, but the way mathematics was taught at that time certainly was. Indeed, it was a great pleasure to solve problems in elementary geometry, with this wonderful toolbox we had. We had to write a problem every week and there were always questions that needed a real effort. This had already changed a lot when my sons were at the secondary school, which is certainly better adapted now to train the whole population, but much less adapted to future mathematicians.

When I entered the university to study mathematics, I was confronted with the best mathematicians of that time, who were also the best possible teachers: François Bruhat, Jacques-Louis Lions, Henri Cartan, Laurent Schwartz, Jacques Deny..., and also Pierre Samuel who gave complementary lectures at the Ecole de Sèvres. At that time I had already decided to become a mathematician, even if I did not know exactly what it meant. Under the influence of my brother, I also studied computer science and even worked a few weeks in a company. But I finally opted for pure mathematics, and analysis in particular, and went to Orsay, where Jean-Pierre Kahane welcomed any newcomer cheerfully, even if at that time the seminar of harmonic analysis was overcrowded. He had succeeded in creating in Orsay one of the best centers in Harmonic Analysis, with visitors from all over the world. I had the chance to listen to the course of Elias Stein on Singular Integrals in 1966–67. It was a great experience. At the same time I followed a course of Yves Meyer, who was then a very young professor. He proposed a list of questions to me, which I was lucky enough to solve with his help, and I was the first one to defend a thesis under his supervision. He was a great supervisor for me. After a few years (I was 26), all my dreams had been fulfilled as far as mathematics were concerned.

If I just stop here, I would leave the reader with the feeling that all was simple and easy.

But life is long, and proofs have to be given over and over, in particular of the fact that one is a mathematician. I always suffered from shyness, which did not recede with participation in conferences. Giving a talk was at first a nightmare for me (the first talk on my work was given by Yves Meyer). Also, I had married in between, had a first son at 28, then a second one five years later. Even if I loved mathematics, the birth of my sons was an incomparable experience, which fulfilled my life. I also had left the research position I had at the CNRS (Centre national de la recherche scientifique) to become professor at the University of Orléans. I would have said at that time that I did not feel like enough of a mathematician to stay on at the CNRS.

The decision to take up a position in Orléans could have been disastrous. My family lived in the Paris area, and I commuted between the two towns during my entire career. In fact I never regretted my decision. The University of Orléans is a small one, compared to the University of Paris-Sud. But I felt at ease there. The mathematics department was fairly recent and I liked the challenge of creating there a scientific atmosphere. With time I started to like teaching. Moreover, all my mathematical activity was mixed with training younger mathematicians, which I could do there.

When choosing mathematics, I had made a double mistake. One concerned me: I thought of myself as someone lonely, not really good at human relations. The other one concerned mathematical activity, which I thought of as a lonely one. I was wrong on mathematics, and I was wrong on me. I realized rapidly that I had to thank mathematics to have given me friends from all over the world. During the last thirty years, I have worked with many colleagues or students of different nationalities. With age, I have also learnt that there is place for many kinds of mathematicians and mathematics.

Ten years ago, I was at a conference in Africa with a small group of French mathematicians. On our way back to France, we found ourselves spending a couple of hours at the airport, a little anxious that our flight be in time, which provides just the right atmosphere to speak of life in general. One of us asked the other ones: "What profession do you give for yourself when you are asked?" I most probably mumbled that I said I was a teacher (maybe a typically French answer, where one likes the feeling of belonging to a large family, such as the one that goes from teachers at elementary schools to university professors). He then told us: "I say that I am a mathematician." And then I realized that yes, I could say the same, and in a way it was the best definition of myself, at least professionally speaking.

22

Why I am a Mathematician

John P. D'Angelo

Dedicated to everyone who enjoys mathematical thinking

Nearly all mathematicians have been asked "why are you a mathematician?" Our answers have varied according to our ages, our moods, and who asked the question. The answer "I love math so much that I have no choice" satisfies some people, but I wonder whether an audience of other mathematicians will accept it. Nonetheless I offer the same answer here, but I disguise, clutter, and embellish it with examples.

Most of my research involves complex analysis in several variables [D1]. I therefore sometimes reply to "why are you a mathematician?" with "the answer depends on several complex variables." No one variable gives the full story, but all contribute to it. I'll start with a few childhood memories and continue to follow my love affair with math chronologically.

As a child I found and enjoyed clever ways to do arithmetic. For example, to find 83×78 I imagined $83 \times 78 = 78 + 82 \times 78 = 78 + (80+2)(80-2) = 78 + 6400 - 4 = 6474$. I knew that $1 + 3 + 5 + \cdots + (2n-1) = n^2$, because I could count an n-by-n array of dots by looking at what one might call the *principal minors*. There is one dot in the first minor, three more then fill the second minor, five more fill the third, and so on. Such delightful tricks were ubiquitous and compelling.

Baseball also helped me to think mathematically. I stared for hours at statistics in the newspaper and on baseball cards. As a result I knew the three-place decimal expansion of every rational number with denominator up to 20. For example, $\frac{3}{13}$ was a batting average of .231, or $\frac{11}{17}$ was a winning percentage of .653. Some of my fun with baseball was more mathematically subtle, such as regarding it (as did many other fans) as a Markov Chain [D3]. I loved trying to think clearly about numerical data and baseball provided many numbers and a context for discussing them.

Dept. of Mathematics, Univ. of Illinois, 1409 W. Green St., Urbana IL 61801 *E-mail address*: jpda@math.uiuc.edu

As a child, I followed the (then hapless) Philadelphia Phillies. One particular situation arose often and galled me: early in a close game, two out, man on second base, eighth-place hitter for the other team coming up. Invariably the manager would walk the hitter intentionally. The next batter (the opposing pitcher) would make an out, and the radio announcers would laud the managerial move. One inning later the opposition would score. The announcers would conveniently forget the previous inning. I was sure that better strategy would be to pitch to the eight-place hitter and get him out; the next inning would then start with a nearly automatic out. The book [TP] obtains this conclusion and clearly analyzes many more situations. Baseball again provided a place for mathematical thinking.

I had the following problem for homework in ninth grade.

Exercise. A plane travels one way at a rate of 380 miles per hour and returns the same distance at a rate of 420 miles per hour. What is the average rate for the trip?

I solved the problem for arbitrary rates r_1 and r_2, discovering the harmonic mean and an obvious inequality it satisfies. Here $399 < 400$. Everyone who has taken a round-trip flight knows that one spends noticeably more time on the slower leg, and hence it matters more.

How do we average rates? Given the pair of rates (r_1, r_2), we first take reciprocals, obtaining $(\frac{1}{r_1}, \frac{1}{r_2})$. Then we average these numbers, obtaining

$$\frac{\frac{1}{r_1} + \frac{1}{r_2}}{2}.$$

Then we take the reciprocal again, obtaining the *harmonic average*

$$\frac{2}{\frac{1}{r_1} + \frac{1}{r_2}} = \frac{2r_1 r_2}{r_1 + r_2}.$$

The following abstract idea summarizes the solution. To do something (find the harmonic average), first change perspective (take reciprocals), then do a simple version (find the arithmetic average), and finally change perspectives back (take the reciprocal). This method has been a recurring dream in my love of mathematics.

Let's consider more examples, in (more or less) increasing order of sophistication.

- For x a real number, put $f(x) = a(x - b) + b$. What is the nth iterate of f? Using T_b to denote translation by b and M_a to denote multiplication by a, we see that $f = T_b M_a (T_b)^{-1}$, and hence the nth iterate g is given by $T_b M_{a^n} T_{-b}$, or by $g(x) = a^n(x - b) + b$.
- How do we express a linear map $L : V \to W$ between finite-dimensional spaces as a matrix M? We use bases to identify $\mathbf{F}^m$ with V and $\mathbf{F}^n$ with W. If the isomorphisms are ϕ and ψ, then $L = \psi M \phi^{-1}$.
- How do we exponentiate a matrix A? We write $A = PDP^{-1}$ (hoping that D is diagonal) and then $e^A = Pe^D P^{-1}$. (Here of course we need to delve deeper and study the Jordan canonical form.)

- For $0 < \alpha < 1$, what is $(\frac{d}{dx})^\alpha$? The Fourier transformation $\mathcal{F}$ diagonalizes differentiation. We naturally write

$$\left(\frac{d}{dx}\right)^\alpha f = \mathcal{F}^{-1} M_{(i\xi)^\alpha} \mathcal{F}(f).$$

In other words, we take the Fourier transform, multiply by $(i\xi)^\alpha$, and undo the Fourier transform.
- What is the definition of a smooth map between manifolds?
- What is a CR mapping?
- When are two holomorphic mappings between balls *spherically equivalent*?

All these questions come down to the abstract philosophy indicated above. One does something by first changing perspective, then doing some simple related thing, and finally changing back. I sometimes tell my students, only partly kidding, that one solves all *word problems* the same way: convert the words to math, let the math do the work for you, and then convert the answer back to words.

As a college student I struggled with surface integrals and flux. Every surface integral seemed either to be 0 or 4π. I didn't get it. One day, in a rigorous course on calculus of several variables, the professor (Frank Warner) said "Now you know the precise meaning of those dx^i's and $\frac{\partial}{\partial x^i}$s." Suddenly the sun broke through; I had no choice but to keep studying math until I did know their true meaning.

I took a differential geometry course my senior year. On the first day we were asked to write our names, list what math courses we had taken, and indicate what we wanted to be when we grew up. For the first time I wrote "mathematician."

Complex variables grabbed me. I have made a career of regarding z and its conjugate $\bar{z}$ as independent variables! I saw the Cauchy integral formula as the two-dimensional analogue of the flux integrals I had not understood before. Furthermore, this magical formula expressed an arbitrary holomorphic function as a superposition of geometric series. Wow! Around this time I started to learn how to think geometrically, although I verified assertions via computations.

As a graduate student I started to become a mathematician. I read lots of math for the first time, but that wasn't my style. I wanted to do examples and computations before I did proofs. After some floundering around, I started working with Joe Kohn. He was the most fantastic role model imaginable. Even though he was department chairman at the time, he always had time to talk about $\bar{\partial}$. At the blackboard he would make a big mess of L^2 estimates and then suddenly say "work out the damned thing yourself." Trying to do so helped me become a mathematician. All I wanted then was to tell Kohn something he didn't know about $\bar{\partial}$. Five years after I got my Ph.D. I finally succeeded in doing so.

I remain a mathematician because basic questions in several complex variables and CR geometry have been implanted in my brain. I affirm my love for mathematics by briefly describing something that has intrigued me for years [D2].

The complex numbers are not an ordered field, and hence it seems impossible to prove inequalities about holomorphic functions. If such a function f (defined on a connected set) is real-valued, then it must be constant. If analysis studies inequalities, then how on earth can one do complex analysis? The naive answer is to prove inequalities on $|f|$ or on Re(f) and Im(f). *Hermitian symmetry* provides a stunning abstraction that includes and clarifies such inequalities. For example let Ω be a domain in $\mathbf{C}^n$ that is closed under conjugation. Consider a holomorphic function $R : \Omega \times \Omega \to \mathbf{C}$ satisfying the Hermitian symmetry condition

$$R(z, \overline{w}) = \overline{R(w, \overline{z})}.$$

In this case $z \to R(z, \overline{z})$ is real-valued, and it is thus possible to prove inequalities about it. Taking $R(z, \overline{w}) = \frac{f(z) + \overline{f(w)}}{2}$ recovers the real part of f, and taking $R(z, \overline{w}) = f(z)\overline{f(w)}$ recovers $|f|^2$. Taking more general R has led me to many ideas in several complex variables, including a developing theory of complexity in CR geometry. I am a mathematician because I want to know everything about Hermitian symmetry. I am a mathematician because I love math so much that I have no choice.

References

[D1] J. D'Angelo, *Several Complex Variables and the Geometry of Real Hypersurfaces*, CRC Press, Boca Raton, FL, 1993.

[D2] J. D'Angelo, *Inequalities from Complex Analysis*, Carus Mathematical Monograph No. 28, Mathematics Association of America, Washington, D.C., 2002.

[D3] J. D'Angelo, Baseball and Markov Chains; power hitting and power series, *Notices of the American Math. Society*, April, 2010.

[TP] John Thorne and Pete Palmer, *The Hidden Game of Baseball*, Doubleday and Company, Inc., Garden City, New York, 1984.

23

Why I am a Mathematician

Robert E. Greene

We think we are in charge of our own lives. But psychology would have it that in actuality we are driven by subconscious forces not only beyond our conscious control but outside our full comprehension. We can modify these forces by rational control. But at heart we are carried along by currents over which we have limited influence.

This was surely so in my life. My parents assured me later that, at a time too early for me now to remember, well before my fourth birthday when I began formal musical training and long before I had engaged in any mathematical study whatever, I was showing interest in two main things: music and numbers. The interest in music is easy to understand: my parents were professional musicians and I was surrounded by music from the beginning. And I demanded to be taught to read music long before I evinced any interest in learning to read words. But the source of my early fascination with numbers is less obvious. Many of the people in my more extended family had been in technical professions, engineering in particular. But mathematics did not figure in my immediate family environment at all.

Even so, mathematics seized my imagination long before I was in a position to know anything about it in a formal way. Mathematical patterns of thought are of course the common property of everyone to some extent. They seem to be innate in the human mind. For me, and early on, this particular form of thought was one I wanted for my own above all others.

Some success later on in mathematical competitions encouraged me to think that I would be able to pursue the subject successfully as a profession. And being a university professor seemed to me in any case a natural choice of occupation, since it was my father's. But the point of origination was simply that mathematics seemed to me from the very start how I wanted to think.

These two interests, mathematics and music, both arising early in my life, never quite allowed the one to triumph over the other. Only in my early twenties when a practical choice was a necessity did I choose mathematics over music as my way to make a living. But music remains with me. And the concerts of the chamber orchestra in which I play today seem as much professional events to me as a mathematics conference.

Mathematics as a career turned out to suit me in ways that go beyond the fact that I found and still find the subject of compulsive interest. I am a naturally solitary person (although as it happens most of my papers have been joint!), and spending a great deal of time thinking alone about topics that I choose myself seems to me an ideal job. I would not have prospered being told what to do nor having to work in groups. Academic mathematics is demanding work if one does it seriously, but the demands are self-imposed and imposed alone. It has seemed to me outrageous good fortune that I have been able to make a living doing exactly what I would have done with my time had I been born so wealthy that I did not have to do anything to earn a living.

Another aspect of academic life that has been important to me is its permanence. It was the fashion when I was growing up in the 1950s to look down on "security" and the desire for it. But in fact, the security of tenured academic employment has been important to me. I am a person who likes things not to change, and the permanent nature of my employment—I have worked at UCLA for going on forty years—has been very satisfying.

When it comes to that, the permanence of mathematics itself is also something that has meant a lot to me. I would not have cared for a scientific field where things moved so rapidly as they have in biology in recent decades. Of course such fast motion also brings large opportunities to be present at the creation, as it were, and to participate in starting revolutions. But the relatively slower pace of mathematics has pleased me more. One of my first serious papers was the solution in 1969 (jointly with H. H. Wu) of a problem brought up explicitly by Hilbert and Cohn-Vossen in 1932 in their *Anschauliche Geometrie* (the question of the rigidity of punctured closed surfaces of positive Gauss curvature). This pace, with more than thirty years from explicit proposal to solution, suited me fine and still does. I like to see mathematics make progress of course, as we all do. But I am happy that revolutions do not occur every year.

It would not be correct to say that mathematics has always seemed a perfect profession to me. As in any human activity, human weakness plays a role. One of the striking parts of Hawking's *A Brief History of Time* is the passage where he describes how others have tried to steal his work. And, if people would stoop to do this to such a man as Hawking, not surprisingly people have tried it on occasion with me, as with many others. And, after all, modern mathematics began with a priority fight (between Newton and Leibniz). Mathematics as a subject is perfect. As a political activity, it often illustrates the well-known dictum that academic politics is so bitter because so little is at stake. But of course it does not seem so little when the stakes involved are one's own. (Ironically, the remark on the bitterness of academic politics, often attributed to Henry Kissinger, is itself of disputed attribution.)

Still, looking back, there is no other job that I can imagine having liked as much as academic mathematics. In the end, what other people have done to one or even what they might think of one's work fades into insignificance in the face of the amazing opportunity to spend one's entire adult life in secure employment thinking about exactly what one would have wished to think about had it been an absolutely free choice. As I near an age where retirement might be appropriate, with, however, no intention whatever of retiring as long as my present good health continues, I look back with the absolute assurance that I took the right road in my choice of profession—no doubt not the right road for everyone but most surely the right one for me.

24

Why I am a Mathematician

Jenny Harrison

My earliest mathematical memory was a joyful one. The two candles on my birthday cake were a thing of beauty. As I blew out the tiny flames, I cherished that doubling of the previous year's singleton, and loved two as twice one from then on. My mother's candle geometry continued into my late teens, each display giving us a moment of shared pleasure before the pattern melted into the chocolate icing. I still associate moments of mathematical insight with the taste of fine chocolate.

When I was five, and for the next ten years, I became an explorer of the woods, a pathfinder. The woods were my haven, my creative playground. Trails and streams seemed so intimate with the land. Where did they go, where did they come from? I felt an intense longing to follow a stream wherever it led me. The stream voiced both chaos and determinism in its flow. Its mysterious source spoke of implied foundations. Unknown paths were a metaphor of life and mind, and they took me on journeys to new terrains. I was always looking for shortcuts, minimal paths to new elements. There were no words, just concepts and relationships in the woods. I learned fluid dynamics from streams, geometry from the turn of a leaf, biology from crayfish and tadpoles, and mechanics from branches blowing in the wind. Sometimes I would sit for hours in a favorite tree and think about numbers or geometrical relationships. School was a distraction from the woods.

When I was in the first grade, a group of adults came to my school to observe me. They asked me many questions. One man asked me to divide a circle into five parts. I drew a nice big beautifully symmetrical circle and made the clean divisions. Then he asked me to divide it into eleven parts. The pie slices were exactly the right size. I saw the pleasure on the faces of the adults, and enjoyed sharing the simple geometry with them.

My principal believed that accelerating bright students harmed their social development, so I was kept with my age group. To stave off boredom, I devised projects to do in all my classes. I taught myself everything; reading, arithmetic, and later, the piano. In the third grade, I learned how to take square roots, and figured out an algorithm for finding cube roots. I believed I could figure anything out for myself, given enough time.

In high school, when my high school geometry teacher was hospitalized for a semester, I was asked to teach solid geometry and did. We could not afford a slide rule and in my science classes I had to work out all of my calculations with a pencil and paper. I was fast and accurate and could always beat the other students with slide rules when we played racing games with long division. In my senior year I participated in a competition that required solving half a dozen problems at home in a week. I loved thinking about these problems without time pressure. The feeling of pleasure when I solved each one was similar to the joy I felt from the candle geometry or drawing the circle and its slices. A judge called to say I had found novel, concise solutions and thought I was a natural born mathematician. My trig teacher said it was a sin I was giving up mathematics for music. But I did not know where mathematics went. The only trail to higher mathematics I had found led me to a book, *Men of Mathematics* by Eric Temple Bell and I felt unwelcome. If I had known that mathematics was primarily about structure as opposed to calculations of numbers, nothing could have kept me away.

One day, when I was twelve, I turned the light on in the dining room and experienced a sense of awe on the nature of light. For an instant, I saw not just rays, as those streaming from the afternoon sun, but the light seemed to fill the room in a wave-like motion. I could sense some kind of duality between waves and rays which I had never heard anyone speak about. I was so amazed that I turned the light switch on again and off again, and asked aloud, "What is this?" I could never see it so clearly again, but I knew that something wonderful and mysterious was hidden there and I wanted to understand it. I told my brother what I had seen. He drew me a picture of the two-slit experiment, explaining how light behaved sometimes like a particle and sometimes like a wave. Which was it? Was it some combination? I vowed to try to figure this out. A few years later, I picked up my brother's book on relativity and felt completely baffled. For the life of me, I could not understand that train speeding by the station with its strangely altered time!

I had to wait until I was a senior in high school in Tuscaloosa, Alabama to take a physics class. Little did I know, but the teacher refused to let girls in his class. He could not do this legally, but he had his ways. After the first homework assignment on vectors, I asked him a question. I remember the question and his response to this day. He had on a dark suit with a thin tie and neatly coiffed black hair. He was standing on the other side of his wooden desk, and said, "If you do not know the answer to that, you don't belong here." From that moment on I internalized his perception that I could not understand physics. After all, Einstein's relativity had flummoxed me, and light had been such a mystery; his viewpoint was consistent with my experience. It was only when some decades later I accidentally discovered some new results in physics through pure mathematics that I realized that the teacher had been wrong. I expect I would have become a physicist if it had not been for that teacher.

At age fifteen, instead of sitting in high branches to contemplate, I turned inward and began a practice of daily meditation at home. The explorations in the woods gave way to explorations of layers of consciousness. I became a teacher of yoga and the habit of mindfulness was set for life and directed my mathematical thinking to come later. In later years, I learned how to sleep on problems and solve them accurately and without forgetting, all without waking up. I did this every night and continue to do so, alternating active thinking with passive listening, using these two modes in a daily rhythm, like pushing a child on a swing higher and higher.

Also at age fifteen I discovered music and the piano. A kind neighbor let me play her piano for a year while I worked my way through John Thompson's five books *Modern Piano Method*. While *Men of Mathematics* discouraged me from mathematics, the allure of Bach and Beethoven was clear and I entered the University of Alabama as a music major. At age twenty, boredom set in from the long hours tied to the keyboard, and I could not seem to overcome terrible stage fright. I decided to quit.

With only one more year of college, my thoughts gravitated to three things I had not been able to figure out—the nature of light, the nature of time, and the nature of consciousness. I no longer trusted experts; philosophers and gurus had stated as irrefutable fact things that made no sense; physicists were always devising theories and backing off from them. I wanted to understand for myself, and believed now that only mathematics offered a path I could trust and that would never let me down. So, in my senior year I took my first math course beyond high school trigonometry, a one-variable calculus course. I memorized it all and got my A, but was left utterly confused by the thousand seemingly unrelated facts in the cold methods book. It was not a good beginning. I wrote an essay about two viewpoints of space that I perceived, coordinates vs geometrical, to try and sort out the mess. In the fall of 1970, I was awarded the first Marshall Scholarship for a student at the University of Alabama. My winning essay was on the nature of consciousness. The Marshall Commission sent me to the University of Warwick where my formal mathematics education began under the tutelage of Sir Christopher Zeeman.

At that time, even though I did not know any mathematics to speak of, I knew a great deal about how to build neural nets in my mind associated to the layers and structures of mathematics. Zeeman interviewed me when I arrived in the fall of 1971 and was struck that I knew less than the first week of their undergraduate course on algebra, their first month of analysis, and no geometry, but had read Kelley's book on point set topology. His friend, an Alabama math professor Bob Plunkett, had written him a letter about me. Though I had planned to take the three year math undergraduate course in two years, Zeeman told me to take it in one, and not worry about prerequisites. In retrospect, this might not have been good advice, but I did it and made the top scores on the final exams that year.

I moved into the graduate program. Zeeman took me on as his student and showed me many things on his slate blackboard with colored chalk. He would give me interesting problems to work on. For the first time in my life, I had a mentor, someone to talk to. Two months in, he included the following problem in his assignment. If there is a surface with triple branches spanning a curve, is there an embedded surface spanning the same curve? He did not tell me this was then a famous unsolved problem, closely related to the even more famous Plateau's soap film problem. It was a Friday afternoon, and I worked hard all weekend trying to figure it out. On Monday at lunch I saw him sitting across the room with a group of faculty. He shouted out, "Did you solve the problem?" I was dismayed because I had gotten nowhere. Everyone laughed at Zeeman's joke, and so did I, once I understood, but I was also flattered that he had thought I might have been able to see how to do it and I never totally stopped thinking about the problem.

In the fall of 1974, my last year at Warwick, Zeeman and I were having lunch and he wrote Stokes' theorem on a paper napkin in very black ink with a beautiful Mont Blanc fountain pen. I had never heard of differential forms, much less this theorem. He said, "This is the most beautiful theorem in the whole world. See how the 'd' upstairs goes downstairs and becomes a 'del?' Analysis and geometry are linked together!" My path into mathematics was set for I never stopped thinking about this, either. I did not know that I would one day find a way to understand both Plateau's problem and Stokes' theorem in one theory unifying analysis and geometry, but that is what happened.

After I left Warwick, I had the good fortune and privilege of finding other great teachers in geometric analysis and differential topology: Whitney, Sullivan, Milnor, and Hirsch. They encouraged me to chart my own course in mathematics, and it is to them that I owe whatever success I have enjoyed. But this is the story of how it all began.

25

Why I Became a Mathematician

Rodolfo H. Torres

Perhaps I could say that I was always attracted to math because of its precise structure and order, its beautiful patterns, its challenging puzzles, its intellectual abstraction, its intriguing mystery, and even its mysticism. I could say that I decided to do math when I saw I enjoyed studying and exploring it and immodestly started to think I had a talent for it. I could say that I fell in love with math when I discovered it can efficiently describe symmetries I could observe in the world I lived in and reveal hidden ones in universes I did not even knew existed. Or maybe I became a mathematician simply because math is tantalizingly addictive; once you experience the moment in which you solve that problem that kept you awake at night, all the frustration and exhaustion of the hard work just melt away and you are invaded by a sublime feeling that you want to experience again. But, the truth is that I do not exactly know when I became fully aware of all this.

I certainly did what one has to do to become a mathematician. I studied hard, learned from others, and tried to be creative. But, growing up, I was a down-to-earth guy who enjoyed simple and mundane things too, many of which would take precedence over my then neglected passion for mathematics. Particular circumstances and people that crossed my life opened and closed doors for me, facilitated or limited my choices, and were responsible for my eventual decision to try to become a mathematician. One day I realized I had chosen mathematics and would be with it in sickness and in health (literally). Hopefully I will live happily ever after with it since, as in the fairy tales, love (the romantic kind) also played a role in my choice for mathematics.

Let me start with the first person that stimulated my passion for the subject: my grandma, *abuela Monona*. My grandma was a very educated person even though she did not finish high school. For a woman born in Argentina in 1888, getting to high school was quite an accomplishment, since most girls were then sent home to help with other "more important" matters of life after a few years of school. Monona could speak several languages, was a good musician and painter, and was extremely skilful with handcrafts too. These things were then "expected" of educated girls, but she was also very good at math. I remember her telling me how some of her teachers begged her father to let her go on with her studies, but her father did not let her; what a waste of talent.

She taught me my first numbers and how to count, and started to teach me card games as soon as I started to talk. My first mathematical "discovery" took place while talking about numbers with her when I was in the early years of elementary school. I realized one day the simple fact that adding two even numbers or two odd ones produces an even number but adding an even to an odd number produces an odd one. I was amazed about this arithmetic wonder but nobody except grandma paid any attention to me. She was not only fascinated with what I told her but also encouraged me to think about the similar situation when sum is replaced by multiplication. That was the beginning of a very fruitful collaboration (through which I made my first buck with math). I continued to share with her all my discoveries and what I was learning at school, and she continued to quiz me and provoke my thoughts with further questions.

Realizing that I was fast with computations, had a good counting memory, and could follow complicated steps, grandma kept teaching me various card games. Soon she was having me as partner playing against some other old lady friends of hers. She liked to brag about her little grandchild and bring me in to play when they were visiting. I was thrilled with the "one peso per game" we used to make, which we split 50/50. Abuela Monona was always proud of me and built my self-esteem more than any other person in my life. Having high self-esteem is very helpful in mathematics when confronting frustrating and hard-to-solve problems.

Professional mathematician, however, was not a real profession that anybody at home could think of at that time. My grandma always told me I should become something important like a doctor (the real kind, not the Ph.D. type). My parents were always very encouraging in my education too and laid it out very clearly from an early age that I should follow a university career. I did not feel any pressure and they were always very supportive, but I think they had in mind some of the more common professions like engineering, accounting, and such. Later in life my father was very happy and proud to see that I enjoy what I do, but I remember him telling me in my youth that math was not a real job; it was something that "engineers did on the side." Despite this he also shared with me mathematical curiosity and we enjoyed solving puzzles together and computing odds in games. Like his own father who had degrees in chemistry and law, my father had many intellectual interests that he tried to pass on to me. Unfortunately he could never finish his medical or his engineering studies. Perhaps because of this, I switched at some point from my grandma's choice of medicine to my dad's preference for engineering.

I had a great math teacher in elementary school who mentored me a lot (and defended me from other teachers when I misbehaved in their boring classes). Math was certainly my favorite subject, but I also liked science, history, geography and writing. It was the era of the space race and the moon landing, so like many other kids I also thought of becoming an astronaut. When I finished seventh grade my mother, a tireless and very determined woman from whom I probably inherited a lot of stamina (another good thing to have to do math), gave me only two choices. I could go to one of the two eighth-twelfth grade magnet schools in my hometown run by the National University of Rosario. One of them, the Polytechnic Institute, focused on math and the physical sciences and was associated with the School of

Mathematics, Physics, and Engineering at the university. I could not imagine at that time that choosing it would seal my professional future and much of my personal life too.

At the institute I learned a lot and enjoyed working on some of the endless math or science projects we had to do. I started to think that I would become a researcher of some sort. Nuclear physics or electrical engineering became my choices, though math was still my favorite subject. I had some wonderful math teachers and I learned through them that people could really become mathematicians. They were also the first to suggest I should do such a thing, but the other careers seemed more fashionable.

I made some of my best friends and had some great classmates with whom I engaged in very stimulating academic interactions. Among these classmates there was a very intelligent and beautiful girl, Estela Gavosto, with whom I ended up sharing math and my life. My growing attraction to mathematics was nothing compared to my attraction to her. But I was more successful with math than with her. She did pay some attention to me, we were friends and passionately argued about every possible subject, but I tried to gain her heart over and over again without much apparent luck. Conquering her heart proved to be more difficult than any theorem I ever encountered. It eventually took me more than ten years to accomplish it and it is by far my best result.

While I was a good student, there were still many healthy distractions (beside Estela) that did not let me focus more in my studies. Hanging out with friends was one of them. Playing rugby in a club was another. I enjoyed rugby very much and dedicated a lot of time to it; during military times there was not much else to do besides sports in terms of entertainment. In the sporadic democratic periods we had I become a student representative, idealist, and interested in "justice." I thought of law or a political career. Then the military returned to power and my mother thought being a student representative was dangerous. She told me to stop being a representative or she would move me to another school. That was probably the only serious argument we really had in our lives, but in the end the choice was clear. She probably saved my future in more ways than one, perhaps my life too.

And so it was that I arrived at college claiming I will study nuclear physics or engineering, but neglecting my true passion for math; hoping to keep up my parallel athletic activities without seeing I could not run so fast anymore, and trying to convince myself that I was no longer in love with Estela (who was, of course, in the same university and had already decided to study math). In summary, I was totally in denial. But I got an unwanted full year to think about my future because I had to do the mandatory military service. Math was my companion there too, as I kept myself awake during the night guard shifts thinking about math problems.

In those days math, physics, and all the engineering careers at the university in Rosario had almost the same first two years of required courses. I could study for a few math classes while in the service but not for physics because I could not attend the labs. In this way I kept up with math but not with other courses. When I returned to full-time study I did not enjoy physics as much as I did in the past, while math was becoming more and more amazing to me. I had to make a choice. If I were to study physics I would have to lose one year to catch up. At that time that sounded like an eternity for a subject I was starting to have doubts about. On the other hand I could continue with math courses right away, I would have again some terrific professors that I had already met and, guess what, I would get to be a classmate of Estela in every single class. It was another clear choice to be made.

By the last year of my degree I was certain I wanted to have a future in academia. I was already teaching and realized I enjoyed it a lot. I was dating Estela, and my rugby career had come to an end. No more distractions, and a desire to pursue a Ph.D. in mathematics became apparent. Democracy had returned to the country and with it many previously emigrated mathematicians. The final push came from some of them, who told us we should do a Ph.D. in the U.S. and helped us get scholarships to do so. The rest of the story is common to most mathematicians: some great teachers, a thesis, postdoctoral positions, a tenure-track job, promotions, etc.—things more appropriate for a vitae than for this type of note—so I will spare the reader those boring details.

I still wish I could work for NASA some day and help in a space expedition to some distant planet, but the idea of being in a space vessel now gives me claustrophobic feelings. I am pretty sure I will never be an astronaut. I still think I could have been a lawyer but my kids make fun of that and, with respect to a career in politics, I had more than enough with the year I served as Faculty Senate President at my University. I still dream of becoming a fiction writer but I doubt I have the talent. I try to go often to the gym but I always have to watch for old injuries so no more intense sports for me. I kept some professional interest in other subjects (for example I have done collaborations with biologists in what are really physical questions) but in the end I always return to my math problems.

All along mathematics was there for me, seducing me but patiently waiting until I could give it my full intellectual attention. Mathematics's beauty and challenges that started to fascinate me long ago are still the same. Mathematics continues to be bread for my intellectual appetite while it literally puts bread on the table for my family (so I guess it has become a real job). It has given me some great collaborators and wonderful students and even kept my mind busy and blocked away depressing thoughts when I was troubled with illness. I cannot separate my path to mathematics from my journey in life with my beloved wife and I think that sometimes things happen for a reason. I am thankful to all those that guided me into this profession and helped me make the right choice.

So why did I become a mathematician? I am not sure I have fully explained it, but given a chance to relive my life I would choose mathematics all over again.